TYPE-B
CYTOCHROMES:
Sensors and Switches

TYPE-B CYTOCHROMES:
Sensors and Switches

Johnathan L. Kiel

United States Air Force
Air Force Materiel Command
Human Systems Center
Armstrong Laboratory
Radiofrequency Radiation Division
Biotechnology Branch
Brooks Air Force Base, Texas

CRC Press
Taylor & Francis Group
Boca Raton London New York

CRC Press is an imprint of the
Taylor & Francis Group, an **informa** business

First published 1995 by CRC Press
Taylor & Francis Group
6000 Broken Sound Parkway NW, Suite 300
Boca Raton, FL 33487-2742

Reissued 2018 by CRC Press

© 1995 by CRC Press, Inc.
CRC Press is an imprint of Taylor & Francis Group, an Informa business

No claim to original U.S. Government works

Library of Congress Cataloging-in-Publication Data

Kiel, Johnathan L.
    Type B cytochromes : sensors and switches / Johnathan L. Kiel.
        p.   cm.
    Includes bibliographical references and index.
    ISBN 0-8493-4576-6
    1. Cytochrome b.   2. Cellular signal transduction.   3. Cell metabolism.   I. Title.
QP671 .C78K54   1995
574.87'61—dc20   94-36841

A Library of Congress record exists under LC control number: 94036841

Publisher's Note
The publisher has gone to great lengths to ensure the quality of this reprint but points out that some imperfections in the original copies may be apparent.

Disclaimer
The publisher has made every effort to trace copyright holders and welcomes correspondence from those they have been unable to contact.

ISBN 13: 978-1-315-89831-5 (hbk)
ISBN 13: 978-1-351-07741-5 (ebk)

Visit the Taylor & Francis Web site at http://www.taylorandfrancis.com and the
CRC Press Web site at http://www.crcpress.com

PREFACE

When I proposed this book to CRC Press on July 22, 1991, the connections among feeder cell and thiol support of lymphocytes *in vitro,* green hemoproteins of red blood cells and the liver, NADPH oxidase of granulocytes and macrophages, nitric oxide synthase of leukocytes and neurons, and nitrogen oxide reductases of plants and bacteria were very tenuous. The feeder cell phenomenon was considered totally replaceable by exogenous simple nonphysiologic thiols. This requirement was seen as an *in vitro* artifact resulting from nonphysiologic levels of oxygen and its toxic products in cell culture media. Yet experiments with exogenous oxidase and peroxidase preparations *in vivo* showed that energetic redox processes that did not generate ATP were not the haphazard, nonspecific injurious processes seen *in vitro.* These nonspecific oxidative killing mechanisms used against cancer cells and infectious agents by granulocytes and macrophages seemed to be inexplicably linked to development of a specific immune response. The discovery of the redox-mediated effects of tumor necrosis factor (which were very similar if not identical to those of bacterial lipopolysaccharide and gamma-interferon) suggested a close redox cooperation between macrophages and lymphocytes regulated by cytokines. In addition, the discovery of nitric oxide synthase activity as essential to the physiologic responses of vasculature, the brain, and insulin-producing beta cells of the pancreas indicated that this specialized NADPH oxidase played roles well beyond the one of wholesale killing by phagocytes.

In my letter to CRC Press, I stated that it was time to draw together the information on electron transport chains and the central role and repeating theme played in them by type-b cytochromes. I further stated that nitric oxide synthase was at the same place NADPH oxidase was 12 years ago. I proposed to show evidence that predicted a type-b cytochrome or similar hemoprotein was required for the second messenger or neurotransmitter function of nitric oxide synthase. On July 28, 1992, White and Marletta reported in *Biochemistry* (31, 6627, 1992) that nitric oxide synthase was the first catalytically self-sufficient mammalian cytochrome P450 (type-b cytochrome), containing both reductase and heme domains on the same polypeptide. This discovery connected the most important redox second messenger, cell-to-cell communicating enzyme with type-b cytochrome biochemistry. It mechanistically connected nonlethal specific cellular responses to nonspecific redox agent effects. Overall, this book attempts to portray type-b cytochromes as primordial proteins that have played and continue to play a central role in the evolution and functions of metabolism, cell growth, and survival in bacteria, plants, animals, and man.

THE AUTHOR

Johnathan L. Kiel, D.V.M., Ph.D., received his B.S. degree in 1973 and his D.V.M. in 1974 (both summa cum laude) from Texas A & M University. He received his interdisciplinary Ph.D. in biochemistry and microbiology from Texas Tech University Health Sciences Center, School of Medicine, in 1981. He is a diplomate of the American College of Veterinary Microbiologists (since 1984) and served on the National Examination Committee for the College from 1988–1991.

Dr. Kiel is the Chief of the Biotechnology Branch, Radiofrequency Radiation Division, Occupational and Environmental Health Directorate of the United States Air Force Armstrong Laboratory. He is author of over 50 scientific publications and holder of ten patents with four more pending. He received the Otis O. Benson Award for the Greatest Scientific Contribution to the U.S.A.F. School of Aerospace Medicine in 1984 and 1989, the Air Force Systems Command Science and Technology Award in 1991, the Research and Development 100 Award for one of the most technologically significant new products of the year 1991 (the Quantitative Luminescence Imaging System), the Harry G. Armstrong Scientific Excellence Award for outstanding research in occupational and environmental health, and the Air Force Association Scientist of the Year Award (Texas Branch) for 1994. He is the United States Air Force's 1994 recipient of the Basic Research Award for his research in mechanisms of interaction of biological systems with nonionizing electromagnetic radiation.

CONTENTS

Dedicated to the memory of
Lawrence Philip Kiel
(March 10, 1964 - February 14, 1993)

ACKNOWLEDGMENTS

This book was prepared with support provided in part by the United States Air Force Office of Scientific Research, the Strategic Environmental Research and Development Program, and by the Cooperative Research and Development Agreement No. 92-262-2A between USAF Armstrong Laboratory, and CRC Press, Inc. Many thanks also to Ms. Arlene M. Schirmer (AL/DOSP), who illustrated the book.

Chapter 1

GENERAL PROPERTIES OF A PRIMORDIAL CELLULAR SENSOR AND SWITCH

CONTENTS

I. BACKGROUND

A. THE ORIGINAL CELLULAR SENSOR/SWITCH

The main theme of this book is that a molecular sensor/switch exists in cells that allows them to effectively respond to life-threatening nonspecific environmental factors. A corollary to this central theme is that the archetype of this molecular sensor/switch and the metabolic pathways in which it was incorporated evolved and diversified into general cellular energy-generating and defensive pathways. In other words, because of the antiquity and essential nature of the sensor/switch motif, any candidate protein or family of proteins found today should hold pivotal roles in a large variety of seemingly unrelated metabolic pathways.

What was the environment like in which this molecular sensor/switch first appeared 4 billion years ago when life emerged on Earth? The atmosphere was predominantly carbon dioxide, nitrogen, and water vapor (in that order of

1

abundance).[1,2] Volcanic and electrical activity and ultraviolet radiation from the sun probably led to the production of methane, ammonia, hydrogen, carbon monoxide, nitrogen oxides, sulfur oxides, hydrogen sulfide, and volatile acids in the atmosphere. The surface of the earth probably contained clays, mineral acids, transition metals such as manganese, iron, and copper and their oxides, alkali metals such as sodium and potassium, and other metallic ions such as those of magnesium, calcium, and zinc.[3]

The atmosphere and surface of the earth were probably permeated with radiation from radioisotopes in surface rocks and released by volcanic activity and meteoric impacts, intense ultraviolet radiation from the sun, and heavy-particle cosmic radiation. Intense heat and shifting magnetic and electrical fields from geologic and atmospheric changes were probably also present. Mars and Venus retain some of these conditions today. Both have atmospheres predominantly made up of carbon dioxide. On Venus, the original greenhouse effect found on Mars, Earth, and Venus at their origins has been carried to the extreme.[1] On Mars, which has lost most of its carbon dioxide atmosphere to washing out by rain and freezing into the polar caps, the surface shows highly oxidative activity driven by ultraviolet light and catalyzed by iron salts and oxides.[1,2]

Electrical, volcanic, and meteoric activity in the early earth's atmosphere may have driven the spontaneous formation of amino acids. The surface clays surrounded by an anaerobic reducing atmosphere could have formed the reactive surfaces for the formation of carbon and nitrogen compounds such as porphyrins, nucleotides and nucleic acids, and primitive proteinoids.[3,4] Such a primordial proteinoid has been successfully formed on clays in such a simulated primitive atmosphere.[4] The proteinoid formed around heme (see Figure 1) and demonstrated peroxidase activity.[4]

Peroxidases are more like reactive reduction/oxidation surfaces with more potential substrates than actual Michaelis-Menten enzymes.[5-21] Therefore, a reactive redox protein could have appeared early in biochemical evolution without the benefit of nucleic acids and could have mediated reduction and oxidation reactions among transition metal ions, atmospheric gases, mineral acids, and water. Such a primitive hemeproteinoid could have trapped and utilized the products of water radiolysis—hydrated electrons, hydroxyl radical, and hydrogen peroxide.[22-24] Certainly, such reactions could have been directly destructive to the newly formed hemeproteinoids.[22] However, if they trapped various reductants, the oxidizing potential of the radiolysis products would have been dissipated. Photochemical reactions driven by ultraviolet radiation not only would have yielded products directly from simple chemicals and gases, but would have also interacted with porphyrin proteinoids in which the transition metal ion was replaced with magnesium, calcium, or zinc ions or the metal-binding site remained unoccupied.[25-27] The photoreactivity of the porphyrin would have been dissipated by coupling transition metal ions, such as those of iron, copper, cobalt, or manganese, to the porphyrin center.

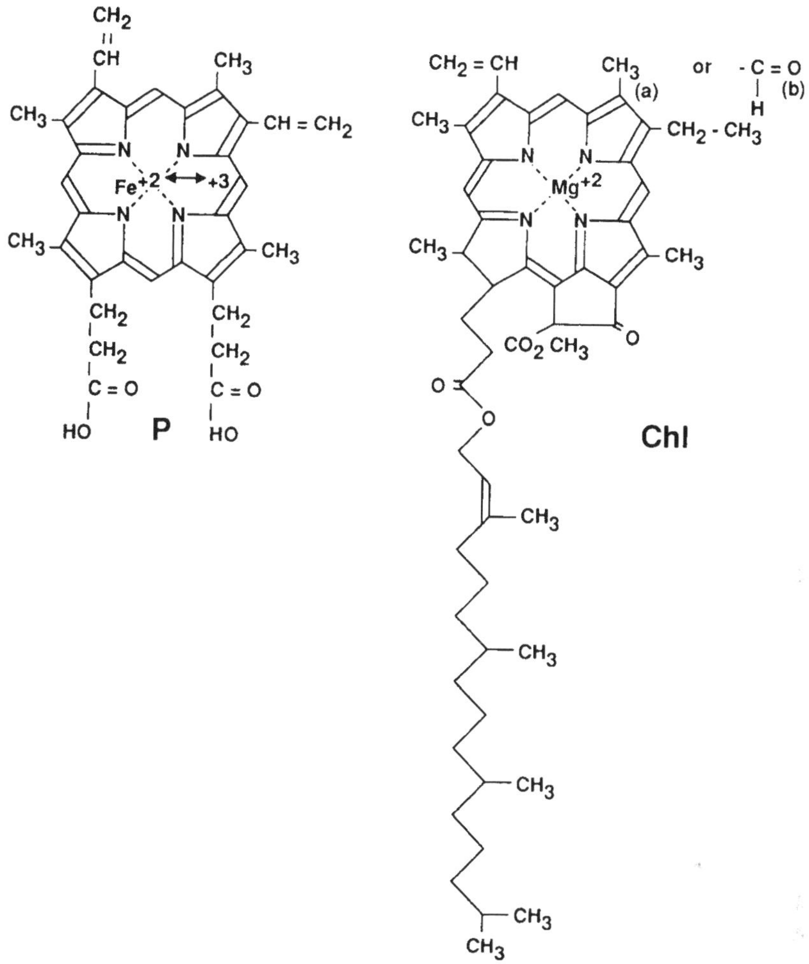

FIGURE 1. Comparison of protoheme (P) structure with that of chlorophyll (Chl: a and b).

Since such a porphyrin proteinoid would have first appeared at phase interfaces (such as clays with liquid water coatings), their hydrophobic porphyrins and external hydrophobic amino acid residues would have enwrapped the porphyrin, and their external hydrophilic amino acid residues would have built an interface with water. This process would have eventually led to the globulization and release of the proteinoids into an aqueous environment.[28] The first heme or metal porphyrin proteinoids would have had the porphyrin noncovalently linked because of hydrophobic binding driven by an external coat of water and the rarity of oxygen that would have made the formation on hydrocarbons of hydroxyl groups, carbonyl groups, disulfide linkages, and other oxidation-driven cross-linking agents difficult. However, some oxygen formation was driven by the breakdown of water and carbon dioxide in the atmosphere by strong ultraviolet light.[2]

Therefore, the predominantly reducing atmosphere would have favored compounds with metallic ions bound to charge-transfer and coordinating compounds like the porphyrins and aromatic and cyclic amino acids such as tyrosine, histidine, and tryptophan. Also, amino acids that were favored by a strong-reducing environment and the availability of sulfur from volcanic activity would have been potentially abundant. These too could bind to transition metals as well as form a complex through the metal with the porphyrins. Cysteine would have been such a sulfur-containing amino acid.

The suggestion recently made that ribonucleic acids could have been the first enzymes and self-replicating molecules is compatible with the concept of hemeproteinoids emerging as the metabolic reactive surface that led to archetypal microorganism metabolism.[29-33] The Achilles' heel of ribonucleic acids is their susceptibility to oxidation at the glycolic hydroxyl moieties of the ribose sugar. The biochemical evolution that was necessary to stabilize nucleic acids was the reduction of the C-2 position of the sugar leading to the deoxyribonucleic acids. This process could have been easily mediated by a heme or porphyrin proteinoid.

Therefore, ribonucleic acids that could have best effected an association between themselves and the right heme or porphyrin proteinoids could have best survived the severe reduction/oxidation environment driven by heat and radiation of the early earth's atmosphere and surface. Some associations were probably deadly in that complexes of magnesium, calcium, or zinc with porphyrin proteinoids or porphyrin/clay complexes catalyzed reductive and oxidative destruction of ribozymes and "ribonucleosomes" or led to, at least, base damage and mutation. Such redox reactions could have been so powerfully directed that they could split water, releasing hydrogen and oxygen. Therefore, the extremely active redox biochemistry of porphyrins and hemeproteinoids and their derivatives (chlorins and isoprenes) was a double-edged sword that could cut or heal.

This original nucleotide/porphyrin protein mutualism for survival could have potentially led to the nicotinamide adenine dinucleotide, flavin mononucleotide, and flavin adenine dinucleotide and adenosine triphosphate-binding and utilization seen in redox chains or fused miniredox chains (a single protein/nucleotide-dependent oxidoreductase). This association could also be envisioned as leading to protein synthesis by RNA because of the propensity for binding certain parts of the proteinoid to certain sequences of nucleotide residues. This process would have favored protection of certain sequences of polynucleotide and condensation of proteinoid subunits (RNA replacing the original clays, which facilitated the formation of both polymers).

Other biochemical polymers with high redox and photochemical activity that may have been contemporary with early porphyrin proteinoids, hemeproteinoids, and ribonucleic acids are melanins and humic acids. Aromatic amino acids, nitroaromatics, and polyhydroxyphenols that trapped hydroxyl radicals and nitrogen oxides formed by radiolysis and electric discharges could have formed autoxidative polymeric products. These products prior to polymerization could react with transition metal ions, hemeproteinoids, and ribonucleic acids, protecting or destroying them or participating with them in various futile redox reactions.

Therefore, the candidate for the primordial molecular sensor/switch of cells should have had the following properties: (1) interaction with radiolytic and photolytic environmental products, like hydrogen peroxide, hydroxyl radicals, hydrated electrons, and other organic and inorganic free radicals; (2) reactivity with primitive atmospheric gases like carbon dioxide, carbon monoxide, sulfur oxides, hydrogen halides, hydrogen, nitrogen oxides, methane, and other hydrocarbons; (3) interaction with nucleotides and nucleic acids; (4) interaction with metallic ions, especially those of transition metals; (5) interaction with phenolic and other aromatic compounds (i.e., nucleotides, aromatic amino acids, melanins, and hydroquinones); (6) sensitivity to oxygen when the sensor/switch was in the reduced state; (7) integrability in hydrophobic microenvironments, but great sensitivity to pH changes, especially gradients; and (8) an ability to modulate, dissipate, or direct redox potential obtained from radiolytic or photochemical reactions.

It soon becomes apparent that these properties are most likely to be found in a class or family of proteins (or analogous structures) rather than in a single species of protein because of the antiquity of the sensor. The type-b cytochromes are presented here as the modern representative of the candidate for the original cellular sensor/switch.

B. HISTORY AND CLASSIFICATION OF TYPE-B CYTOCHROMES

Cytochrome b's were discovered by H. Fischer during the period of 1910 to 1940.[34] They were categorized by D. Keilin based upon their visible light-absorption spectra.[34] They are classified based upon their longest wavelength light-absorption bands, the alpha bands, wavelengths between 500 and 600 nm, and their heme type. Most cytochrome b's are imbedded in membranes by post-translational modification. A few are made water soluble by proteolysis. Examples of both types are listed in Table 1.

The insoluble cytochrome b's such as NADPH oxidase may be associated with solubilizing and stabilizing subunits or chaperonins as are all cytochrome b's to be transported into cellular membranes.[35-39] As shall be shown in later examples, the cytochrome b's appear to be almost uniformly associated with flavoproteins or fused segments of polypeptides that bind flavins and/or nonheme transition metal ions. The cytochrome b's, in spite of diversity of function and primary structure, show chemical relatedness.[20,39,40] As shall be seen when

TABLE 1
Solubility of Type-b Cytochromes

Type	Soluble	Insoluble (Membrane-bound)
Cytochrome b_5	+ (RBC)	+ (L)
Cytochrome P450	~ (+*)	+ (L)
NADPH oxidase	~	+ (PMN, M)
NO synthase*	+ (PMN, M, P, L, B, E, GI)	+ (E, M)
NAD(P)H diaphorase	~	+ (B, P, L, PMN, M)
Nitrate reductase	+ (HP, F, Ba, BGA)	+ (Ba)
Nitric oxide reductase	~	+ (Ba)
Mitochondrial (Complexes I, II, III)	~	+ (HP, Y, Pr, A)
Chloroplast and bacterial photosystems	~	+ (HP, BGA, Pr)

Note: RBC, red blood cells; L, liver; (+*), NO synthase is considered a soluble cytochrome P450; PMN, polymorphonuclear leukocytes; M, macrophages; P, pancreas; B = brain; E = endothelium; GI = gastrointestinal tract; HP, higher plants; F = fungi; Ba = bacteria; BGA = blue-green algae; Y, yeasts; Pr, protozoa; A, animals.

examples are examined and compared, the specific functions are more dependent on quaternary associations and compartmentalization than structural differences of cytochrome b's.

Is there any evidence that this modern class of heme proteins was first associated with RNA or DNA? Evidence is found in the requirement of cytochrome b's to be associate' with thiolated transition metal-binding polypeptides. Another enzyme critical to DNA protection from oxidative damage, endonuclease III, has a nonheme iron polypeptide unit that allows it to bind to DNA to fulfill its function as a repair endonuclease.[41] Could this be the remnant of the original association of cytochrome b's with nucleic acids because they have a high affinity for thiolated transition metals, or are these structures analogous but of different origins? Other circumstantial evidence for this nucleic acid/cytochrome association is found in that (1) the plant hormone indoleacetic acid is bound to RNA (transfer RNA, messenger RNA, and ribosomes) by the peroxidizing action of horseradish peroxidase, a cytochrome b-like hemoprotein;[42] (2) horseradish peroxidase requires a bound calcium ion for optimal activity;[43,44] and (3) horseradish peroxidase is naturally found bound to DNA in plants.[45]

Other evidence for the association of hemoproteins with cytochrome b-like properties with nucleic acids and gene expression include the inactivation of estradiol by binding of the hormone to proteins. This covalent linkage is mediated by uterine peroxidase associated with an NADPH oxidase in the endoplasmic reticulum of endometrial cells.[46-48] The activation of ferritin messenger RNA destruction by heme but not by iron ions (a nonredox effect) suggests post-transcriptional regulation of mRNA by heme and, therefore, originally by a hemoprotein that supplied the heme.[49] Also, free heme or hemin upregulates heme oxygenase activity (expression) as does oxidative stress.[50-52] Beginning with the general properties of type-b cytochromes and ending with specific examples of systems involving the type-b cytochromes, the hypothesis that the primordial environmental sensor/switch of the first living organisms was a type-b cytochrome will be supported.

II. CHEMICAL AND PHYSICAL PROPERTIES

A. CHARACTERISTICS OF THE HEME CENTER

The heme of type-b cytochromes contains ferrous or ferric iron in a proto-porphyrin IX. The structure is illustrated in Figure 1. This heme is buried in a hydrophobic pocket within the cytochrome. It is noncovalently linked in this pocket with the fifth ligand of the iron being the side chain of a histidine residue and the sixth ligand being water, thiolate, or a quinone. [12,34,53-56]

The type-b cytochromes may contain one or two such hemes or form polymeric types of the cytochrome.[57,58] Most of the type-b cytochromes are membrane bound and have four transmembrane domains.[58] Those that are cytoplasmic have quaternary structure, with other polypeptides associated with them noncovalently or fused to them covalently that allow for the hydrophobic portions to be transported in aqueous medium.[59,60] Examples of the latter are NADPH oxidase of leukocytes and nitric oxide synthase of brain, endothelial cells, macrophages, neutrophils, and liver.[59,60] Examples of the intramembrane cytochrome b's are found in P450, b_5, and nitrate and nitric oxide reductases of bacteria.[34,58,61]

The type-b cytochromes can be considered anionic peroxidases, although they generally show weak peroxidase activity and then self-bleach.[62,63] Bleaching may result from an abortive compound II that yields hydroxyl radical and destroys the heme.[62] Most are of a molecular ratio of 21 to 23 kilodaltons (kD), although they may be larger when incorporated in a gene-fusion product or polymeric form.[64]

B. STRUCTURE AND FEATURES COMMON
WITH OTHER HEMOPROTEINS

The type-b cytochromes are related to peroxidase and the globlins—hemoglobins and myoglobin.[65-67] However, these proteins are usually cationic. Horseradish peroxidase, lactoperoxidase, myeloperoxidase, eosinophilic per-oxidase, hemoglobin, and myoglobin all contain one or more noncovalently linked protoporphyrin IX-type hemes.[5,6,8,10-12,20,34,67] Lactoperoxidase, like type-b cytochromes, binds phenolic compounds such as tyrosine more tightly when reduced by thiol.[68] Other peroxidases react readily through redox mechanisms with phenols and thiols.[17,18,21,47,48] These peroxidases and globins, like type-b cytochromes, bind a variety of anions such as halides, nitrates, nitrites, cyanides, superoxide, and gases such as nitric oxide, oxygen, and carbon monoxide, depending on the oxidation state of the iron ion.[19,54,65,69]

C. REACTIVITY WITH OXYGEN AND FREE RADICAL PRODUCTS

Autoxidation of type-b cytochromes appears to be a reasonable mechanism for superoxide and hydrogen peroxide and, subsequently, hydroxyl free radical formation by these hemoproteins. However, in only three type-b cytochrome redox chains is the b cytochrome considered the terminal oxidase—erthrocytic green hemoprotein, NADPH oxidase and nitric oxide synthase.[59,60,62,63] In mitochondria and chloroplasts, when superoxide is formed, it is at the hydroquinone, not the cytochrome b's.[71-73] Even the reduction of nitrate to nitrite by nitrate reductase is mediated, not at the cytochrome b level, but through a molybdenum-containing polypeptide.[74] Actually, in these latter three systems the type-b cytochromes are probably protected by their intramembrane locations and generally anaerobic environment. In other words, they are in a strongly reducing environment. The usual immediate electron transport couples with these cytochromes are nonheme iron sulfur proteins, transition metal complexes, and/or hydroquinones or quinonoids.[34,35,56,72] This indirect interaction with oxygen is not unprecedented. In the oxidation of NAD(P)H by horseradish peroxidase, the NAD(P)· free radical is formed by the peroxidase and is the immediate source of the electron for reduction of oxygen to superoxide.[16] Similarly in damaged chloroplasts and mitochondria, the superoxide comes from quinone radicals.[71,73] Therefore, type-b cytochromes are expected to be directly reduced or oxidized by reductants or oxidants, respectively, other than oxygen. Unlike peroxidases, they do not first have to be oxidized by hydrogen peroxide to start the autoxidation cycle of free radical/peroxide production.[16-18] Type-b cytochromes appear to have some catalase activity, forming oxygen bound to the reduced heme iron from two hydrogen peroxide molecules.[63] This oxygen may be released as superoxide or molecular oxygen.[63] Catalase resembles type-b cytochromes in that it is part of a truncated redox chain. It contains bound NADH. Also, like cytochrome b's, it is a poor peroxidase with a K_m for peroxide of about 1 M, and it can form a compound II-like intermediate. The latter can peroxidize lipids suggesting that it behaves like a heme-hydroxyl radical complex.[75] This general preference for oxidants other than oxygen follows the probable early emergence of type-b cytochromes in biochemical evolution prior to the presence of oxygen in the earth's atmosphere.

D. REACTIVITY WITH THIOLS, PHENOLICS, AND QUINONES

Thyroxine, phenols, and quinones mediate the oxidation of NAD(P)H by horseradish peroxidase.[16-18] These co-factors are similar to those found in the redox chains and in close association with type-b cytochromes.[34,35,56,76,77] Accompanying and sometimes replacing hydroxylated aromatics in this role are thiols.[34] They not only protect the cytochromes from autoxidative self-destruction, but they also are essential to their normal functions.[54,62-64] These include even the truncated redox chains of nitric oxide synthase and methemoglobin superoxide transferase (GHP).[63,78] NADPH oxidase seems to be the only type-b cytochrome redox chain lacking direct evidence for such a required association with thiols and/or quinones.[79-81] The point remains controversial.[59]

E. PH AND THERMAL SENSITIVITY

The instability of type-b cytochrome systems is evident during purification of the respective type-b hemoproteins from nitric oxide synthase, NADPH oxidase, nitrate reductase, and green hemoprotein.[58-60,63,64,74,79] The yields are progressively lower with each step when the cytochrome iron ion is in the reduced state.[64] Although the type-b cytochromes are anionic, and those such as in NADPH oxidase, mitochondria, and chloroplasts operate *in vivo* in acidic environments, during purification, acidic buffers rapidly destroy them.[64,74,80-84] Therefore, the cytochrome b's appear to operate *in vivo* under conditions that should lead to instability but are somehow entropically locked to prevent denaturation. This condition sets the type-b cytochromes up as switches with adjustable sensitivity. How this switch sensitivity is adjusted *in vivo* and how the switching is "read" when flipped is a central theme of this book.

III. GENERAL MECHANISMS

A. IN PHOTOSYNTHESIS

Type-b cytochromes provide the central bridge between photosystems I and II in higher plants.[35,76,83] They are linked to the photosystems by quinonoids or nonheme iron proteins. Chlorophylls are porphyrin-like structures called chlorins (Figure 1) that contain magnesium instead of iron ions and are not contained in protein but rather in membranes.[35] In a sense, they are representative of the primitive state of porphyrins proposed in the first section of this chapter.

The energy collected photochemically by chlorophyll and its satellite pigments (carotenoids, or isoprenoids) is transferred as reducing equivalents through nonheme iron proteins and quinonoids to the cytochromes, which conserve the energy or transfer it to high-energy bonds. The final form of the latter is adenosine triphosphate (ATP). Photochemical energy is recovered from the charged radical separation of the photopigments (chlorophyll becomes a cationic radical) by the formation of anionic quinonoid radicals (from plastaquinone) or thioyl radicals in nonheme iron proteins.[35,76,77] This process translocates or releases protons from the plastaquinone or nonheme iron sulfur protein. When these protons cross the thylakoid membranes of the chloroplasts (from inside to outside), ATP is produced from ADP and phosphate by the intramembrane F_1 ATPase.[35,84-86]

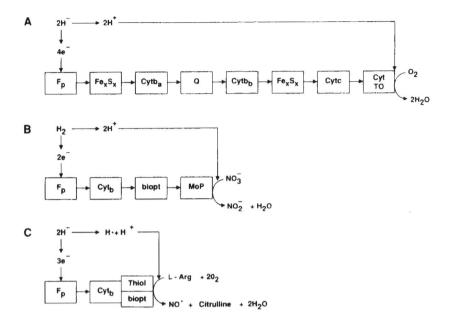

FIGURE 2. Comparison of three redox systems: (A) mitochondrial; (B) nitrate reductase; and (C) nitric oxide synthase. F_p = flavoprotein; $Cytb_{a\ or\ b}$, Cyt_b = various type-b cytochromes; Fe_xS_x = nonheme-iron-sulfur proteins; Q = ubiquinone; Cytc = cytochrome c; Cyt TO = the terminal cytochrome oxidase; biopt = biopterin; MoP = a molybdenum-carrying co-factor; L-Arg = L-arginine.

B. IN MITOCHONDRIA

Mitochondria of plants and animals contain electron transport systems for the conversion of chemical-reducing equivalents into ATP.[34] As in the chloroplast photosynthetic pathways, in mitochondria, ATP-generating sites are immediately before and after cytochrome b's.[34] The transferred electrons in mitochondria come from the splitting of the hydride of NADH into one proton and two electrons (Figure 2). In the case of mitochondria, unlike chloroplasts, the quinone is between the cytochrome b's in the chain, and the nonheme iron sulfur proteins are before and after the cytochrome b pair.[34,35,87,88] The terminal oxidase of the mitochondria is in a position in respect to the cytochrome b's similar to the water-splitting part of photosystem II of the photosynthetic pathway.[34,35] NADH dehydrogenase of mitochondria is in a position similar to that of the anaerobic photosystem I.[34,35] In the mitochondria, NADH is used as the source of hydride instead of NADPH as in photosystem I of plants.[89,90] In eukaryotic cells, NADH is used more as a source of chemical energy (ATP) than is NADPH, which is used more as a source of reducing equivalents for synthetic processes such as fatty acid desaturation, ribonucleotide conversion to deoxyribonucleotide, hydroxylations, and superoxide production.[90] This global look at these two systems shows how they biochemically complement and resemble each other, even when they are found in different species.

C. IN CYTOCHROME P450 SYSTEMS

Cytochrome P450 systems resemble the primordial peroxidase (type-b cytochrome) described in the first section of this chapter.[91] Again, they maintain the same pattern of cytochrome b's seen in mitochondria and chloroplasts. However, they utilize reducing equivalents from either NADH or NADPH through flavin-containing P450 reductases, suggesting a more primitive state than the cytochrome b's of chloroplasts and mitochondria which have become highly selective.[34,35,89,90] In P450 systems, the electrons from NADH or NADPH pass through the flavoprotein to a nonheme iron sulfur protein to the cytochrome P450.[34]

The hydrocarbon to be hydroxylated by the cytochrome P450 to a more soluble material for detoxification and excretion is usually hydrophobic, such as an alkane, alkene, aromatic or polycyclic.[34] The mechanism strongly resembles the oxyperoxidase (compound III) and compound II formation seen in such peroxidases as horseradish peroxidase.[91] In fact, horseradish peroxidase can hydroxylate aromatics and polycyclics, assisting in their epoxidation and arylation of nucleic acids, as seen with cytochrome P450 in the formation of imminent carcinogens.[46,91] Peroxidase can also oxidize long-chain fatty acids and alkenes, as does cytochrome P450.[15] Once again, the conservation of peroxidative and hydroxylating mechanisms of peroxidases and cytochrome b's is seen.

D. IN PHOTOBACTERIA

In the photon-emitting system of photobacteria, reducing equivalents from NAD(P)H are transferred through a flavoprotein to cytochrome b, but here the chain splits and can switch either to cytochrome c for mitochondria-like oxidative phosphorylation or to reduction of the bacterial luciferase (a flavoprotein).[92-94] This luciferase oxidizes long chain fatty acids and transfers the excitons to the flavin of the protein. The excited flavin then irradiates the energy as blue-green light.[95] This reaction is, in a sense, the reverse of photosynthesis.

E. IN METHEMOGLOBIN REDUCTASE

The term methemoglobin reductase is actually a misnomer because the reductase is really a cytochrome b_5 reductase.[96] This short redox chain begins with a flavoprotein that reduces cytochrome b_5 in the cytoplasm of the red blood cell with reducing equivalents from either NADH or NADPH.[96,97] The cytochrome b_5, in turn, reduces the methemoglobin. The cytochrome b_5, normally a membrane-bound cytochrome, as in the liver, is post-translationally modified by proteolysis.[97,98] If this modification fails to occur in development, then a methemoglobinemia results.[97] In this defect, the cytochrome b_5 is not solubilized and, therefore, remains inaccessible to the methemoglobin in the erythrocytic cytoplasm. Whether thiols or nonheme iron sulfur proteins are involved in this redox chain remains to be determined.

IV. SUMMARY

This chapter began with the speculation that a nonspecific cellular sensor/switch began its biochemical evolution with condensation of amino acids on primordial clays. According to this hypothesis, these amino acids coformed with hydrophobic organics and interacted with immobilized transition and other metallic ions. The resulting hemoproteinoids and porphyrin-containing proteinoids were very active in redox chemistry generated by the strong ionizing-radiation and ultraviolet-light environment of the early earth. At this time, ribonucleic acids were also forming but were vulnerable to radiation-induced oxidations and reductions. The binding of the hemoproteinoids to the nucleotides and nucleic acids provided protection from such damage at times and mediated redox destruction of the nucleic acids and nucleotides at other times. So began the nucleotide/organic chemical electron-transfer evolution with type-b-like hemoproteinoids as the speculation suggests.

Modern evidence for this archetypal relationship with the ribonucleotides includes the following: (1) type-b cytochromes are found in many diverse redox systems in bacteria, plants, and animals; (2) type-b cytochromes are usually in the anaerobic end of the redox chains except for a few truncated chains such as NADPH oxidase, NO synthase, cytochrome P45?, and green hemoprotein; (3) type-b cytochromes are very sensitive to pH, hea*, and oxidation in the reduced form when removed from the *in vivo* microenvironment, and they denature easily; (4) these cytochromes have a propensity for interacting with thiols (especially nonheme-iron-chelating thiols) and aromatic compounds, which tend to protect the cytochromes from autoxidation; (5) the cytochrome b's are usually immobilized in membranes that stabilize their structure and allow them to more efficiently participate in energy-capturing and/or -conserving processes; (6) nonheme-iron-thiol proteins and flavoproteins mediate the interactions of type-b cytochromes' with reducing ribonucleotides; (7) endonuclease III, which is involved with repairing oxidative damage to nucleic acids, has a polythiol-iron complex that has no current redox function but is essential for its binding to nucleic acids; and (8) the reduction of the 2-OH on ribonucleotides to decrease their vulnerability to oxidative damage and form deoxynucleic acids is mediated by thioredoxin, a nonheme-iron-sulfur protein that utilizes a stabilized tyrosyl radical and cobalt complex. Therefore, the most basic functional archetypal unit of a type-b-cytochrome sensor/switch should contain a nonheme-iron-thiol complex, reducing nucleotides (NAD(P)H and flavins), a quinonoid, and the type-b cytochrome, itself. This is the fundamental unit that will be examined in this monograph in its various modern forms.

Chapter 2

PRIMITIVE TYPE-B CYTOCHROMES: GREEN HEMOPROTEINS

CONTENTS

I. PHOSPHOENOLPYRUVATE CARBOXYKINASE FERROACTIVATOR

Green hemoproteins (GHP's) are type-b cytochromes found in the brain, pancreas, liver, spleen, and red blood cells that demonstrate ferroactivator activity.[62-64,69,82,99,100-106] GHP transfers reducing equivalents to a hypersensitive disulfide, converting it to thiols, in phosphoenolpyruvate carboxykinase (PEPCK).[102] This process activates the enzymic activity of PEPCK, the conversion of oxaloacetate to phosphoenolpyruvate, the key step for conversion of lactate to glucose (gluconeogenesis).[54,102,104] Thiol/transition metal ion complexes of iron or cobalt can be the source of the reducing equivalents.[106] Whether this is a specific activity of a particular type-b cytochrome or a property of a variety of type-b cytochromes found in the aforementioned organs and tissues remains to be confirmed.

A. IN THE LIVER

In the liver, the ferroactivation of PEPCK is a key activity in the organ's function in gluconeogenesis from lactate produced by anaerobic metabolism, primarily of skeletal muscle.[54,104] Therefore, at least in the liver, this activity is probably the green hemoprotein's primary function. Like other GHP's, the liver GHP shows a low yield in the purification process and rapid heme spectral loss in acid pH buffer.[54] This protein has a molecular weight (M_r) of 21,000 and may exist in a polymeric form.[54] The liver GHP is a cytoplasmic form not an intramembrane protein like cytochrome b_5. Antibody has been made against the rat liver GHP and has been used to demonstrate that it is distinguishable from cytochrome b_5 in liver preparations.[54,99]

B. IN RED BLOOD CELLS

Phosphoenolpyruvate carboxykinase ferroactivator has been found in human and bovine red blood cells in spite of the fact that red blood cells do not contain phosphenolpyruvate carboxykinase.[64,99,101] This observation raises issue with the idea that ferroactivation of PEPCK is the primary function of red blood cell GHP. That green hemoproteins are highly conserved among species is evident based on amino acid composition and on the cross reactivity of antibody against rat liver GHP with human and bovine red blood cell GHP.[64,99,101] This antibody did not cross-react with cytochrome b_5.

The red blood cell GHP has other properties that are very similar to liver GHP.[64,99,101] Red blood cell GHP is a polymeric hemoprotein of $M_r = 82,000$ to 90,000 with a subunit M_r of 22,500. Its heme can also be lost by purification under acidic conditions even though it is an anionic protein. The oxidized red blood cell GHP has visible light absorption peaks at 407 (the Soret band), 507, and 543 nm, and the reduced form has absorption maxima at 419, 549, and 591 nm. Red blood cell GHP is easily bleached by dithionite but shows little change in its visible absorption spectra when treated with hydrogen peroxide alone, unlike true peroxidases like horseradish peroxidase.[62,63] Hydroxyl free radical formation on the GHP has been proposed as the reason for the bleaching.[62]

The bleaching reaction can be considered as the formation of an unstable compound II of the GHP.[62,63,107] Compound II is a one-electron oxidation form of peroxidase that is equivalent to a bound hydroxyl radical.[62,63,107] The following equations illustrate the bleaching reaction:

$$GHP(Fe^{+3}) + 1e^- \rightarrow GHP(Fe^{+2}) \qquad (1)$$

$$GHP(Fe^{+2}) + H_2O_2 + H^+ \rightarrow GHP(Fe^{+3} \cdots OH) + H_2O \qquad (2)$$

$$GHP(Fe^{+3} \cdots OH) \rightarrow GHP_{ox} + HEME_{ox} + H_2O \qquad (3)$$

This set of futile reactions is not the physiologic function of GHP in red cells because no more of the protein can be synthesized in mature red blood cells and such reactions would soon deplete all the GHP in the red cells, which is not the case *in vivo*.[63] This soluble type-b cytochrome has some mechanism of stabilization in red blood cells. As evident during the purification of this cytochrome, the association with hemoglobin seems to be related to this stabilization.[63]

Hemoglobin is a cationic tetramer of two types of hemoprotein subunits (alpha and beta).[66] Upon complete oxidation to methemoglobin, it breaks into monomeric units, becoming more cationic.[63,66,105] GHP readily binds to hemoglobin and especially methemoglobin, making it difficult to purify from red blood cells.[62,63] Hemoglobin undergoes autoxidation when acidified, heated, or treated with high concentrations of various anions (chloride, cyanide, etc.).[75] The activation energy for the autoxidation is 134 kJ/mol (32 kcal/mol).[75] The activation energy for deoxygenation of oxyhemoglobin is 56 to 60 kJ/mol. Therefore, it appears that oxyhemoglobin should deoxygenate before it autoxidizes by a direct superoxide-producing process. If oxyhemoglobin is bound to the cell membrane of red blood cells, the activation energy of deoxygenation rises to 67 kJ/mol.[75] However, this increased activation energy is still below that for superoxide dissociation, indicating that deoxygenation is highly favored over autoxidation to superoxide and methemoglobin.

These facts and the affinity of GHP for hemoglobin and methemoglobin suggest that GHP may mediate or catalyze the autoxidation of hemoglobin and potentially the reverse reaction.[63,107] The following redox process would reverse autoxidation of hemoglobin and spare GHP:

$$2GHP(Fe^{+3}) + AH \rightarrow 2GHP(Fe^{+2}) + A^+ + H^+ \qquad (4)$$

$$GHP(Fe^{+2}) + H_2O_2 \rightarrow GHP(Fe^{+3} \cdots OH) + H_2O \qquad (5)$$

$$GHP(Fe^{+3} \cdots OH) + H_2O_2 \rightarrow GHP(Fe^{+2} O_2) + H^+ + H_2O \qquad (6)$$

$$GHP(Fe^{+3} \cdots O_{\bar{2}}\cdot) + MetHb \leftrightarrow HbO_2 \text{ (monomeric)} + GHP(Fe^{+3}) \qquad (7)$$

$$4HbO_2 \rightarrow (HbO_2)_4 \text{(tetrameric)} \qquad (8)$$

Reactions (5) and (6) represent the catalase activity of GHP, and (7) and (8) represent the superoxide transferase activity. Reactions (4)–(7) represent a complete catalytic cycle. AH represents a reductant such as thiol or NAD(P)H, MetHb is methemoglobin, and Hb is hemoglobin. Therefore, the GHP may serve two vital antioxidant functions—destroying hydrogen peroxide that is a by-product of oxidative metabolism and converting methemoglobin (resulting from autoxidation and chemical oxidation) back to oxyhemoglobin. This process, unlike methemoglobin reductase activity, proceeds readily in a peroxidizing environment, simultaneously decreasing the source of peroxidation, hydrogen peroxide. However, based on the principle of microscopic reversibility, the GHP process presents one danger, that GHP may lower the activation energy of oxyhemoglobin autoxidation and catalyze the production of methemoglobin and superoxide (reaction (7)).[63]

The activation energy of autoxidation of red blood cells has two components—a low-activation-energy component of 18.5 kJ/mol below 29°C and a high-activation-energy component of 53 kJ/mol above 29°C.[75] Again, both these activation energies are well below the activation energy for autoxidation of free oxyhemoglobin (134 kJ/mol). Furthermore, thermally induced oxidative hemolysis of red blood cells was enhanced by wheat-germ-agglutinin-horseradish-peroxidase conjugate, which binds to the cell membrane of the red blood cells.[105] This observation suggests that hydrogen peroxide plays a role in thermally induced autoxidative damage of red blood cells. Contrary to what was expected, the *in situ* thiol-conjugating agent 1-chloro-2,4-dinitrobenzene inhibited thermally induced autoxidative damage to red blood cells.[105] This observation is in accord with GHP's using thiols as reductants to drive superoxide, hydrogen peroxide, and hydroxyl radical production.

C. IN SPLEEN AND OTHER TISSUES

Even though green hemoproteins have also been reported in spleen, pancreas, and brain, there may be other hemoproteins with similar visible light-absorption spectra. Because of the similarity of the spectra of lactoperoxidase and myeloperoxidase to type-b cytochromes, what may be considered to be GHP may, in fact, be revealed to be one of these other hemoproteins upon closer examination.[69] For instance, splenic GHP, or "peroxidase", can oxidize chloride to hypochlorite as does myeloperoxidse from neutrophils and monocytes that may be found in the spleen.[82,100] The pancreas and brain contain many cytochromes including nitric oxide synthase that could be confused with GHP.[108,109]

II. OTHER GREEN HEMOPROTEINS (GHP'S)

A. SNAKE SKIN "PEROXIDASE"

Prior to molting (ecdysis), the upper layers of snakes' skin (stratum corneum) is invaded by eosinophilic granulocytes (heterophils and eosinophils).[110] These cells contain peroxidase and presumably oxidase to produce hydrogen peroxide used by the peroxidase. Ecdysis is a cyclic process involving proliferation and killing of cells of the upper layers of the epidermis.[110] This process also affords some protection from infections that would otherwise penetrate the skin.[110] During the process, lower living layers of the epidermis die in one to two weeks, depositing fluid, pigment, and mature keratin between the remaining living epidermis and the old stratum corneum, which splits away and peels off in a continuous piece or divided pieces. During the phase in which the upper section of the epidermis is dying, the skin and brilles, or spectacles, over the eyes appear bluish or blue-gray (the "blue phase" of molting).[110] The new cuticle surface and underside of the old cuticle is covered with a thiol-rich surfactant-like material containing "peroxidase" activity, measurable by spectrometric (by 2,2′-azino-di-(3-ethylbenzothiazoline sulfonic acid (ABTS) peroxidation), cytotoxicity, and chemiluminescence assays.[110,111] The first assumption was that the "peroxidase" was eosinophilic peroxidase or myeloperoxidase. However, the "peroxidase" activity was rapidly lost upon purification, and even storage at −70°C did not prevent 90% destruction in 24 hours.[110] This peroxidase had catalase-like activity and bleached ABTS when further-purified preparations were tested for peroxidase activity.[110] These are characteristics of an "anionic peroxidase", or green hemoprotein. Also, not all snake species show the eosinophilic or heterophilic influx into the skin during molting. This information suggests that the GHP of snake skin may be endogenous to the tissue and active in the programmed cell death of the epidermal cells.[110]

B. UTERINE OXIDASE/PEROXIDASE

Similar to the molting cycles of snakes, eosinophils influx in rat and mouse uteri (endometrium) during the estrus cycle.[112-114] Estrogen was found to be conjugated to proteins such as serum albumin by the eosinophilic peroxidase.[110,114,115] Therefore, the peroxidase activity in endometrium was considered to be derived from the eosinophils that influxed during estrus. Eventually, however, the peroxidase was found to be synthesized by the endometrial cells themselves along with an NADPH oxidase of the endoplasmic reticulum of these cells.[114-116] The expression of these enzymes was directly controlled by estrogen levels in the uterus.[115] The similarity of this oxidase/peroxidase system to that of the snake skin "peroxidase" is remarkable. Estradiol is an aromatic polycyclic that should be a good substrate for a type-b-like cytochrome. It is likely, although currently unproven, that the uterine oxidase/peroxidase system that inactivates estrogens is a type-b cytochrome system.

III. NONIONIZING ELECTROMAGNETIC FIELD EFFECTS

A. ON ERYTHROCYTIC GREEN HEMOPROTEIN

The simple short redox chain composed of GHP and hemoglobin couple was an excellent test case for type-b cytochromes' being sensors. Microwave radiation (2450 MHz) at intensities ranging from 0.4 to 100 mW/g held at constant temperature showed effects at three levels of red blood cell autoxidation.[63,75,105,107] The assumption was that if a type-b cytochrome system could sense nonthermogenic nonionizing electromagnetic fields, then it would have the exquisite sensitivity required of a highly evolved nonspecific sensor. Thermally induced autoxidative hemolysis of red blood cells was inhibited by microwave radiation at 42°C.[105] No such inhibition was observed at lower or higher temperatures. The autoxidation of the surface of red blood cells was inhibited above 37°C.[75] The higher activation energy component of the biphasic thermally induced autoxidation of red blood cells was not evident during microwave radiation. Finally, purified bovine hemoglobin to which partially purified human GHP was added or fresh human hemolysates showed inhibition of thermally induced oxyhemoglobin autoxidation.[63] When GHP was not present or was allowed to denature before the oxyhemoglobin autoxidation was measured, the microwave radiation inhibition was not observed.[63] Furthermore, the GHP enhanced thermally induced oxyhemoglobin autoxidation in the absence of microwave radiation.[63] The question arises as to how the cytochrome can sense the forces generated by the alternating electromagnetic fields in the face of the substantially greater random molecular forces generated by the thermalization.

B. MECHANISMS OF INTERACTION

The production of superoxide from oxyhemoglobin mediated by GHP can be summarized by the following equilibrium:

$$Hb(Fe^{+2}O_2) \leftrightarrow HB(Fe^{+3} \cdots O_2^{-} \cdot) \tag{9}$$

Even though the GHP lowers the activation energy for this process from 134 kJ/mol for oxyhemoglobin autoxidation alone to 18.5 kJ/mol for temperatures below 29°C and 53 kJ/mol for temperatures above 29°C in its presence, the weak electromagnetic radiation previously described should not be sufficient to alter the response. The electric field component which should favor charge polarization should, therefore, enhance superoxide production. However, if the escape of the superoxide from the hemoglobin is much slower than the frequency of the radiation (2450 MHz), then the field, if intense enough, should pin the charged species (such as superoxide) in place (causing it to vibrate around the equilibrium point of movement). As mentioned earlier, even the binding of oxyhemoglobin to the Band 3 transmembrane anion transport protein does not raise the activation energy of deoxygenation (from about 60 to 67 kJ/mol) enough to make autoxidation of oxyhemoglobin significantly competitive with deoxygenation.[75] The deoxygenation reaction is made less favorable by the binding causing the free energy to change from –2.5 to 5.4 kJ/mol. The final conclusion about these processes is that the dissociation of superoxide from oxyhemoglobin is unlikely to occur without catalysis (lowering of the activation energy). Although GHP enhanced thermally induced autoxidation of oxyhemoglobin as expected of a catalyst, its facilitation of microwave radiation inhibition of the process had no apparent explanation. These results, if generalized, predicted that changing electromagnetic fields of a frequency greater than the escape rate of a bound charged free radical should inhibit charged free radical formation and enhance uncharged radical formation by type-b cytochromes.

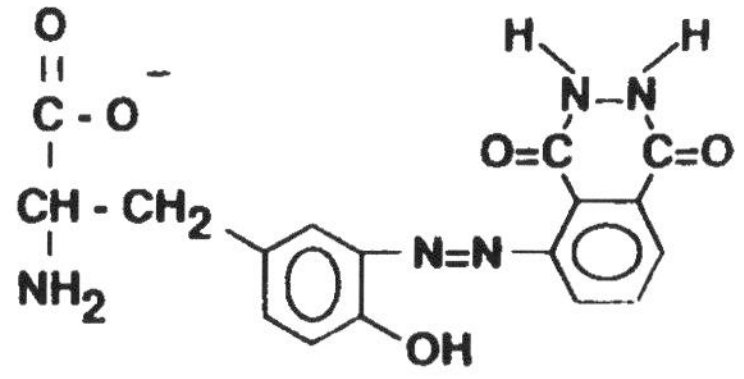

FIGURE 3. The expected diazodimer resulting from diazotization of 3-amino-L-tyrosine and luminol (5-amino-2,3-dihydro-1,4-phthalazinedione).

IV. DIAZOLUMINOMELANIN: SUBSTRATE AND MECHANISTIC PROBE FOR GHP

A. MOLECULAR CONCEPT AND DESIGN

Since type-b cytochromes appear nonfunctional when purified and rapidly denature in the reduced form, a compound was sought that would act as both a redox substrate and a preservative for the cytochrome. Because 3-amino-L-tyrosine (3AT) was previously observed to inhibit peroxidase, the oxidative burst of mouse macrophages, and the autoxidation of red blood cells, it was considered the molecular prototype for a substrate for type-b cytochromes.[117] Luminol (5-amino-2,3-dihydrophthalazine-1,4-dione) has been used as an indicator of redox activity because of its chemiluminescence upon peroxidation.[96,119] However, luminol, in itself, does not bind to hemoproteins, but reacts nonspecifically in a second-order fashion, yielding free radicals that, in turn, interact with free hydrogen peroxide to yield the luminescent species.[119] Thus it is a good indicator, but a poor substrate. Therefore, in order to gain specificity and an indication of the reaction, the combination of 3AT and luminol by diazo coupling was proposed to form a novel substrate.[103,118] Figure 3 shows the anticipated structure.

However, when chemical synthesis of this compound was attempted, a melanin-like material formed.[118,120,121] Based on the similarity of properties of the new material to those of true melanins, the indole-like unit structure of a repeating polymer was proposed, as shown in Figure 4. Because of the highly anionic nature of the black melanin-like polymer, designated diazolumino-melanin (DALM), and its solubility (ability to form a colloidal solution), it was assumed to be largely carboxylated.[120,121]

DALM did react vigorously with erythrocytic GHP.[118] DALM's baseline chemiluminescence in the presence of hydrogen peroxide increased 500 times in the presence of partially purified GHP, but not in the presence of other hemoproteins like horseradish peroxidase or purified hemoglobin. Further studies of the nuclear magnetic resonance (NMR) properties, calorimetry of polymerization, and theoretical quantum calculations in order to predict its color changes as polymerization progressed or pH changed indicated a structure (Figure 5) alternative to that of Figure 4.[120,121] Figure 5 shows an open, less melanin-like head-tail linear polymer. The structure, as depicted in Figure 5,

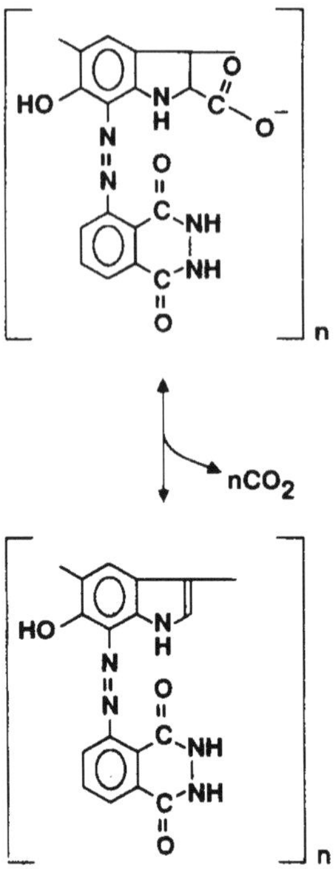

FIGURE 4. The anticipated polymer of diazotized luminol and 3-amino-ʟ-tyrosine.

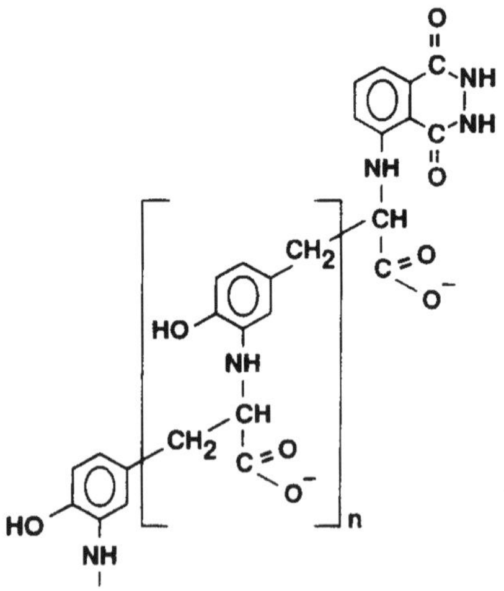

FIGURE 5. The putative structure of diazoluminomelanin based on current physical and chemical evidence.

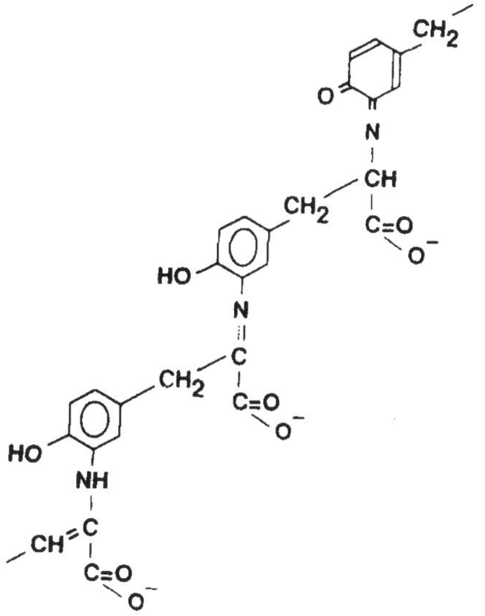

FIGURE 6. Resonance structures of oxidized diazoluminomelanin.

should be nearly white (although a strong ultraviolet light absorber) and tend toward linear fiber formation or monoclinic or triclinic quasi-crystalline formation. Also, the content of luminol (as a chain-capping agent) should be very low, based on this structure. These properties have been observed. As portions of the chain oxidize, forming quinonoidal and isolated conjugated sections, the polymer should become yellow, violet, then burgundy red, brown, and finally, black, dependent on the number of conjugated bonds.[120] Again, these properties have been observed.[118] Another unusual property of this polymer is that polymerization seems to proceed predominantly in the solid state, based on calorimetry and NMR.[120,121] Some of the quinonoidal structures and their isomers are shown in Figure 6. Also. these same studies showed that no diazonium groups are observable in the final polymer, adding more evidence for this specialized open polymer structure. Figure 7 shows how the polymer may shrink, following formation, by internal cyclization into a more melanin-like indolic structure.

The crystalline structure in Figure 8 may form the basis for the solid-state polymerization. The ionic alignment of the diazonium salt (carboxylate/diazonium interactions) allows for displacement of the diazonium group as free nitrogen gas by the alpha amino group of the amino acid subunit. Monoclinic and triclinic crystals of precursors of the fibrous DALM release gas (perhaps nitrogen) in sodium carbonate and hydrogen peroxide solution.[118,120,121] Besides reacting with GHP of red blood cells, DALM was also found to yield enhanced chemiluminescence with the membrane fractions of HL-60 human leukemia

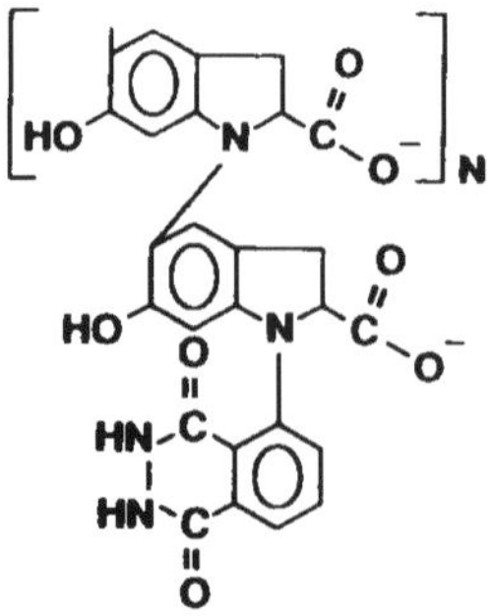

FIGURE 7.   Cyclized (indole tautomer) diazoluminomelanin.

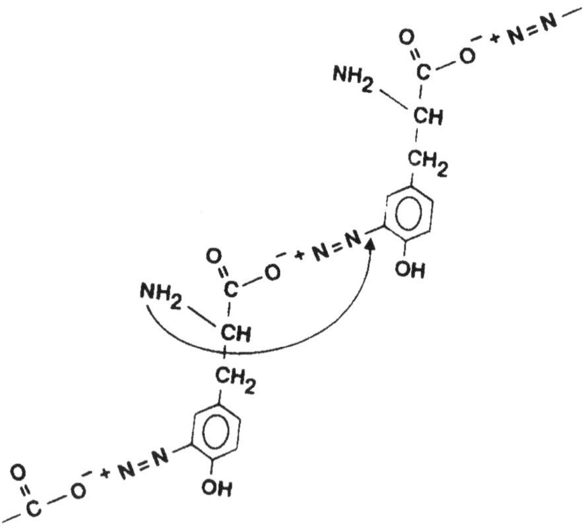

FIGURE 8.   Linear crystal of 3-diazo-L-tyrosine, showing direction of alpha-amino group nucleophilic attack to form diazoluminomelanin.

cells differentiated to produce the membrane component (type-b cytochrome) of NADPH oxidase.[122]

B. PHYSICAL CHEMICAL MECHANISM

What is the chemical basis for the interaction of type-b cytochromes with DALM, and is it related to the normal function of GHP? Do the luminescent properties of DALM give an insight into the *in vivo* reaction mechanisms of type-b cytochromes in general and GHP in particular? The first step in answering these questions was to examine the photochemical properties of DALM. When DALM is treated with sodium carbonate or bicarbonate solution and hydrogen peroxide, it chemiluminesces in a near-steady-state mode for a very long time. A few micrograms can luminesce for in excess of 52 hours without further additions of hydrogen peroxide.[118] Heating the peroxidizing, luminescing DALM solution increases the intensity of the chemiluminescence in an exponential fashion.[118] Only moderate temperatures (25–60°C) and temperature changes were required for significant increases in chemiluminescence.[118,123] The chemiluminescence establishes a steady state when the temperature is held constant.[118]

The thermally sensitive luminescence of DALM can be explained by Type E slow fluorescence.[123-135] The following "forbidden" intersystem crossing of excited quantum states illustrates the slow fluorescence mechanism:

$$S_1 \leftrightarrow T_1(\pm) \tag{10}$$

$$T_1(\pm) \leftrightarrow T_0 + kT \tag{11}$$

where S_1 is the first excited singlet state of a photosensitive component and $T_1(\pm)$ is the charged triplet state at the energy level equivalent to that of the S_1. The $T_1(\pm)$ releases a small amount of energy at kT or less (product of Boltzmann's constant and absolute temperature in Kelvins) to drop to the uncharged triplet state T_0. The T_0 will slowly degrade to the baseline S_0 state by releasing heat or long-wavelength (red) light. The process in reactions (10) and (11) can be summarized as follows:

$$S_0 + Q \leftrightarrow S_1 \leftrightarrow T_1(\pm) \leftrightarrow T_0 + kT \tag{12}$$

where Q is the quantum of light necessary to raise S_0 to S_1.

The intensity of the slow luminescence can be described by the following equation[123,133,134]:

$$I = Ae^{-(E/kT)} \tag{13}$$

where I is the intensity of the slow luminescence; A is the amount of activated state that can potentially be converted from T_0 to S_1; E is the energy gap between the T_0 and S_1 states; k is Boltzmann's constant; and T is the absolute temperature in Kelvins. As the energy gap approaches 0, I approaches the maximum value represented by A. The reason that the slow luminescence does not become fast luminescence is the inherent delay in the forbidden transition (electronic spin flipping) between $T_1(\pm)$ and S_1, even if they are at the same energy level. This barrier is based on the statistical distribution (probability of state) rather than the energy difference.

This process is essential to the capture of photonic energy by chlorophyll.[125,131,135] The chlorophyll becomes a cationic radical upon photonic excitation and transfers its first electron to the Z acceptor in photosystem I, which in turn becomes an anionic radical.[135] Therefore, an excited singlet state (like in DALM) is converted into a charged triplet state (paired radicals, electronic spin doublets), which can provide energy for proton translocation.[135] The proton translocation can eventually provide energy for ATP production by a breakdown of the proton gradient.[35,84] This singlet-to-triplet process is probably also operable in the dissipation of absorbed photonic energy by eumelanins.[136-138] Even in mitochondria, the general mechanism is operable because the hydride (H^-), a singlet state, is passed to flavin, which separates the paired electrons of the singlet state into two electrons of unpaired spin (a triplet) and a proton.[139-142] This process is done through semiquinone formation on the ubiquinones associated with the pair of cytochrome b's in the redox chain.[56]

The increase in DALM luminescence catalyzed by GHP suggests that this cytochrome b facilitates the conversion of triplet state quinonoidal groups to activated singlet states that release photons when returning to the ground state.[123] In other words, the GHP lowers the energy barrier E between the activated singlet and metastable uncharged triplet states. Therefore, the slow luminescence approaches its inherent maximum A in equation (13). This catalysis should increase the sensitivity of the system to electric and magnetic field effects.[123]

In photosynthesis, when the slow fluorescence of chlorophyll is not quenched by electron transfer (ultimately through cytochrome b's), 10 G to 2 kG magnetic fields lead to increased intensity in slow fluorescence.[124,131,134] Magnetic fields of 2 kG or more lead to a decrease in slow fluorescence intensity. Electric fields depositing energy of 4 W or greater also lead to decreased slow fluorescence intensity. The magnetic field effects can be explained by quantum mechanics.[123,124,131] The triplet state degenerates into three vector orientations of equal energy, x, y, and z. However, in the low-intensity magnetic field (less than 2 kG), the orientation where the magnetic field vector is aligned with the external field is at a lower level of energy than the singlet state, and the magnetic field vector that is aligned against the external magnetic field is at a higher level of energy than the singlet state. The T_0, or neutrally aligned state, is at the singlet-state energy level under these conditions. Therefore, the triplet states are no longer energetically equivalent and collapse into a single T_0 triplet state. The equilibrium between the singlet and triplet states is disturbed by the momentary lack of equal probability between these two remaining states. To reinstate the equilibrium in the external magnetic field, the number of singlet-state molecules is increased, and the number of triplet-state molecules is decreased.[123] Therefore, more molecules are available to decay by photon release to the singlet ground state. In high magnetic fields (greater than 2 kG), the singlet state is induced to undergo a phase shift from paired electronic spins to unpaired electronic spins, increasing the total number of triplet-state molecules. In electric fields, the charged triplet-state molecules are favored by alignment with the field and by release of thermal energy in the vibrational cascade to uncharged metastable triplet state.[123] If the triplet energy level is maintained (not allowed to radiate heat or infrared radiation), then the electric field can inhibit the intersystem crossing and facilitate fast luminescence.[123]

Microwave radiation inhibition of thermally induced autoxidation of oxy-hemoglobin catalyzed by GHP is in agreement with this mechanism.[63] When microwave radiation absorption by DALM was enhanced by its immobilization on magnetite nanoparticles (heated faster than free DALM), the slow luminescence was lower for a given temperature than expected.[123] This result is predicted by the conductive salt magnetite reducing the electric fields near the bound DALM in the microwave radiation field to zero, which is in accord with the slow luminescence mechanism.[123] When peroxidizing DALM solutions were exposed to 1-kW, 40-μs pulses of 2450-MHz microwave radiation at one pulse per second, intense pulses of light emission followed the microwave pulses.[118] Again, this effect is consistent with the slow-luminescence mechanism. The electric field of the radiation caused a sudden buildup of triplet-state molecules with a large release of thermal energy in a very small volume, severely and rapidly disturbing the quantum equilibrium. When the microwave pulse terminated, the thermal energy was reabsorbed and the large store of triplet states (represented by A in equation (13)) were converted back to excited singlet states to reestablish the equilibrium of states. This process led to a sudden release of a large number of photons.

V. SUMMARY

Green hemoproteins can be considered primitive cytochrome b's because they are not associated with long membrane-bound redox chains. They are usually associated with a nonheme-iron-thiol complex, the thiol being cysteine or glutathione, and a target protein for electron transfer such as phospho-enolpyruvate carboxykinase or hemoglobin. Other GHP's have anionic peroxidase or catalase-like activity. In the presence of electron donors without an electron acceptor, other than oxygen, reduced GHP's rapidly autoxidize, releasing some superoxide and bleaching with endogenously generated hydroxyl free radical. Based upon the photochemical interactions of erythrocytic GHP with diazoluminomelanin, a slow luminescing quinonoidal polymer, the fundamental chemical mechanism of cytochrome b interactions appears to be catalysis of rapid singlet/triplet state interconversions. This mechanism is probably the basis for proton translocation in ATP-generating cytochrome b systems that utilize quinone/hydroquinone couples in the membrane translocation of electrons and protons.

Chapter 3

NADPH OXIDASE AND ITS TERMINAL CYTOCHROME

CONTENTS

I. STRUCTURE AND ACTIVATION MECHANISM

The purpose of this chapter is to discuss the role of the cytochrome b_{558} in NADPH oxidase. This redox chain system has been thoroughly reviewed elsewhere, and it is not the intent to repeat that process here, but rather to focus on the cytochrome and its role in respect to the entire chain.[143-154] Furthermore, its role as a sensor/switch is more relevant in macrophages and in related long-lived cells than in granulocytes that are terminal suicide killer cells. However, the NADPH oxidase of granulocytes will be briefly addressed in respect to structure and enzymic activity because the redox chain has been so thoroughly investigated in these cells.[59,143]

A. ASSEMBLY OF COMPONENTS

NADPH oxidase is a short redox chain composed of two intramembrane cytochrome components: (1) a hemopolypeptide of about 22 kD with a half-redox potential of –245 mV and (2) a flavoprotein component of about 91 kD.[155-170] The latter component is a stabilizing component for the hemopolypeptide. The two components assemble in sets of two each in the membrane.[59] There are four other components provided by the cytosol for a functional NADPH oxidase: (1) Rac$_2$, a G-protein, (2) Rap1A, (3) P47, a G-activating-protein, and (4) P67.[59,167,169]

Activation of the system following opsonized particulate antigen phagocytosis or gamma-interferon stimulus involves at least two pathways.[171-194] The first is a calcium-ion-dependent one involving a sodium/calcium transporter, calcium-activated phospholipase C, and protein kinase C.[172,176-178,180-189,192] The kinase phosphorylates the cytochrome b_{558} and/or the cytosolic components to complete the NADPH oxidase assembly. Although arachidonic acid release has been associated with NADPH oxidase activation, it is not essential.[181,194] It is associated with the phospholipase activation in the early stages of the activation pathway.[195,196]

Receptor activation of NADPH oxidase is mediated through guanosine triphosphate (GTP) binding proteins (non-ras G-proteins).[169,174] This G-protein activates phospholipase C by releasing its GTP-bound subunit that binds to the phospholipase C.[176] The phospholipase C subsequently cleaves phosphatidylinositol 4,5-biphosphate into inositol-1,4,5-triphosphate and 1,2-diacylglycerol. The latter activates the protein kinase C. Phorbol esters can also activate the protein kinase C, directly, and the NADPH oxidase respiratory burst, subsequently.[195]

In cell-free systems, only arachidonic acid or anionic detergents and the G-protein component are necessary to activate the NADPH oxidase.[196,197] No ATP is required in contrast to activation by protein kinase C. Other ATP-dependent kinases can also activate the oxidase.[169,171,185,186,198] Arachidonic acid ATP-independent, GTP-dependent activation is believed to involve a separate G-protein (of the ras family).[169] Another small subunit (22 kD) of NADPH oxidase binds GTP (the binding site shows homology with the H-, N-, and K-ras proteins); it is rap-1.[173] This protein is probably one of the soluble cytosolic components triggered by arachidonic acid. The other cytosolic component P47 (47-kD subunit) is associated with the transduction pathway between the kinases and NADPH oxidase.[59,175] The P47 subunit is 49% homologous with the ras-GTPase-activating protein (ras-GAP). GAP is believed to stimulate p21ras-GTPase activity. The interaction between ras-GAP and p21ras is weakened by cis-unsaturated fatty acids, specifically, arachidonic acid.[199]

In keeping with the general theme of tightening the hydrophobic/hydrophilic interface of cytochrome b for redox activity, phosphorylation and GTP transfer and binding to the NADPH oxidase should increase its hydrophilicity. The reactivity of NADPH oxidase with hydroquinones and quinone-like compounds such as 1-naphthol and 3-amino-L-tyrosine results in inhibition of the respiratory burst.[117,200,201] Although this inhibition is probably mediated through essential thiols, such thiols are probably at or near the hydrophobic/hydrophilic redox interface (electron transfer point between water-soluble and water-insoluble components).[184] The quinones block the electron transfer path with the hydrophobic ring and probably trap the electron in a lower redox potential well.

B. STABILIZATION OF CYTOCHROME B$_{558}$

The 91-kD component of cytochrome b$_{558}$ has been shown to contain the flavin adenine dinucleotide (FAD) binding site.[170] It is also necessary for the stabilization of the heme-binding of the 22-kD subunit.[57,202] This property has led to confusion as to which of the two subunits actually binds the heme.[170,163] Again, however, such instability of the hemoprotein is characteristic of type-b cytochromes associated with superoxide production.[166,203] It also suggests that the instability is critical to function as well as control. The affinity of NADPH oxidase for anions is evident in the inhibition of its activity by a variety of nucleotides (such as ATP, ADP, or AMP) and anions like fluoride, pyrophosphate, and citrate.[167,172,187,197,204-207] This could also explain its inherent affinity for NADPH over NADH.[208,209]

The 91-kD flavoprotein subunit must be in close proximity to the heme moiety not only for stability but also to transform the singlet pair of electrons transferred to the flavin from NADPH to a triplet pair that is, in turn, transferred to the heme. This conversion (intersystem crossing) is critical to the reduction of the triplet oxygen to form the doublet superoxide.[210] This mechanism is central to the enzymic activity of NADPH oxidase.

This combination of stabilizing properties and functional integrity have been confirmed by the observation that the 22-kD subunit messenger RNA (mRNA) is produced in nonphagocytic cells such as HeLa, HepG$_2$ (hepatic), K562 (erythroleukemic), cultured human endothelial cells, and Epstein-Barr-virus-transformed B lymphocytes.[211] However, stable protein was not observed in these cells. In HL-60 cells, a promyelocytic human leukemia cell that can differentiate into neutrophilic or macrophage-like cell lines upon induction, similar levels of 22-kD subunit mRNA were observed.[211-213] However, upon induction with dimethylformamide treatment, mRNA specific for the 91-kD subunit dramatically increased, leading to a stable, functional, membrane-bound NADPH oxidase.[211,214,215] Furthermore, when membrane components from lymph node and thymus cells were combined with macrophage cytosolic components, a functional NADPH oxidase was reconstituted.[216] Some, but not all, Epstein-Barr-virus-transformed B cells could be activated to produce superoxide by phorbol myristate acetate or by cross-linking immunoglobin on the surfaces of cells that possessed it.[217]

Another observation that links the cytochrome b stability to function is that the differentiation of HL-60 cells into granulocytes or macrophages leads to specific increased luminescence of diazoluminomelanin by the action of isolated membrane components from these differentiated cells.[122] Comparing this result with that using green hemoprotein suggests a strong functional similarity between green hemoprotein of red blood cells and the NADPH oxidase cytochrome b of white blood cells.[146] The constitutive production of the 22-kD hemopolypeptide of NADPH oxidase in such a diverse group of cells could be linked to its being a very sensitive membrane-bound detector of heat, oxidative, and radiation stressors.[218] Such a protein with a low threshold for denaturation, heme release, and autoxidation could be a first-line-of-defense warning to increase heat shock protein production, or antioxidant production. Oxidatively denatured proteins do activate heat shock protein expression.[219,220] Heme release induces heme oxygenase production, a stress-related marker.[51] Furthermore, the heme could act as a receptor for nitric oxide, a neurotransmitter and intercellular communication molecule.

III. FUNCTIONS IN LEUKOCYTES AND OTHER CELL TYPES

A. GRANULOCYTES

1. Neutrophils

Both neutrophils and macrophages demonstrate NADPH oxidase activity initiated by similar stimulants: (1) particulate antigens coated with antibody, such as bacteria or yeast particles; (2) the chemotactic agent N-formyl-L-methionyl-L-leucyl-L-phenylalanine; (3) the cytoskeletal active agent cytochalasin E; (4) the metabolic stimulant sodium fluoride; (5) endotoxin lipopolysaccharide of Gram-negative bacteria; (6) the detergents digitonin, deoxycholate, and saponin; (7) phospholipases C and A_2; (8) antileukocytic antibody; (9) immune aggregates; (10) the plant glycoprotein concanavalin A; (11) complement components C_{5a} and $C_{5,6,7}$ (C-activated serum); (12) the vasoactivating proteolytic enzyme kallikrein, which splits kinins (such as bradykinin) from the globulin of alpha$_2$ globulin; (13) fatty acids; (14) phorbol myristate acetate (PMA); (15) ionophores with their cations calcium and magnesium; (16) lanthium trivalent cations; (17) polystyrene dishes; (18) micropore filters; (19) the plant lectin phytohemagglutinin (PHA); and (20) taurocholic acid.[59,143,146,149,172,183,205,221-226] Leukotriene B_4 and platelet activating factor (PAF) also activate the oxidative burst of neutrophils.[188,227] Therefore, agents that immunologically, chemically, or physically perturb the cell membrane activate the oxidase. The biochemical change that is associated with the physical change in the membrane is the lowering of the K_m for NADPH.[190] The K_m's ranging from 0.05 to 0.03 mM are seen in the neutrophil NADPH oxidase.

NADPH oxidase and nitric oxide (NO) synthase, which are both found in neutrophils and macrophages, are activated in parallel by gamma-interferon and lipopolysaccharide (LPS) or tumor necrosis factor (TNF).[228] The latter two agents enhance the response but gamma-interferon is capable of activating the enzymes by itself.[228-230] That NO and superoxide production can be uncoupled in both neutrophils and macrophages has been clearly shown.[231,232] However, the two redox processes do occur at the expense of each other.[233] Superoxide dismutase enhances NO production, and NO synthase can generate hydrogen peroxide as does NADPH oxidase, although the peroxide is likely to damage the NO synthase.[60,79] Both enzymes require NADPH, oxygen, and at least some thiol as substrates and co-factors.[172] In neutrophils, the NO production can be terminated, but the superoxide production cannot. Evidently, though distinct enzymes, NADPH oxidase and NO synthase are very closely related functionally (parallel evolution) or phylogenetically (common origin).

Superoxide, peroxide, and hydroxyl radical produced as a consequence of NADPH oxidase activity and free transition metal ion interactions kill bacteria, viruses, tumor cells, and some normal tissue cells.[223,235-238] Neutrophils depend on glutathione peroxidase as their primary defense against self-inflicted damage.[145,239] Target cells and microorganisms may undergo induction of defense systems that render this oxidative attack ineffectual.[240-245]

2. Eosinophils and Mast Cells

Eosinophils, like neutrophils, are capable of a vigorous oxidative response to membrane stimulation.[246-248] Upon ingestion of latex particles, eosinophils showed greater glucose carbon-1 oxidation and formate oxidation (a measure of hydrogen peroxide production) than neutrophils, although eosinophils do not phagocytyze bacteria, and therefore, are not as effective at bacterial killing as neutrophils.[192,249] Eosinophils are more likely to release their peroxidase and peroxide into extracellular fluid than neutrophils.[249] This effect was demonstrated by enhancement of iodination by addition of albumin into the cell-culture medium.[247,248] Also, the basic metabolic activity and binding of estrogen to protein through a peroxidase-mediated mechanism is higher in eosinophils than neutrophils, even with the eosinophils' lower rate of phagocytosis. In general, increased metabolic activity of eosinophils has been measured by increased hexose monophosphate shunt activity, glucose oxidation, hydrogen peroxide formation, superoxide anion generation, chemiluminescence, thyroid hormone degradation, iodination, and estrogen binding to protein. These results suggest that metabolic activation (activation of the NADPH oxidase) may occur with cell surface membrane changes without particle ingestion. This property would be compatible with the eosinophils' primary function of combating multicellular or metazoan parasites.[144]

Complexes of IgE with antigen induce the extracellular release of peroxidase from eosinophils, which mediates the killing of target organisms when hydrogen peroxide is also released from the NADPH oxidase in the presence of halides (iodide, chloride, or bromide).[249,250] This process may effectively decrease allergic reactions.[251] However, in contrast to this removal of allergens and subsequent decrease in allergic reactions, eosinophil interactions with mast cells imply an enhancement of such responses.[251,252]

Mast cells have been found to contain peroxidase, which is sensitive to formaldehyde or glutaraldehyde destruction, making it difficult to detect in fixed tissue.[252,253] The existence of an active-transport-type histamine release from mast cells mediated by superoxide or low levels of hydrogen peroxide suggests that eosinophils could mediate or enhance the allergic response by facilitating histamine release by this mechanism.[251,252] Mast cells are induced to release granules when treated with low levels of hydrogen peroxide, eosinophil peroxidase, and a halide.[254] These granules have been shown to be cytotoxic for mouse ascites lymphoma cells (LSTRA) when activated with hydrogen peroxide and halide.[254] The antitumor activity was enhanced by addition of eosinophilic peroxidase, which bound to the granules.[251] Therefore, the extracellular degranulation of eosinophils and mast cells facilitates the host's attack on large parasites like schistosomes and tumor cells and cell masses too large to be phagocytized.

B. MONONUCLEAR LEUKOCYTES
1. Monocytes and Macrophages

As mentioned in the section on neutrophils, the macrophage NADPH oxidase is activated and stimulated by the same materials and processes as the neutrophils. However, the macrophages assume more metabolic and differentiation states than neutrophils. The various states of macrophages show various K_m's of the NADPH oxidase for NADPH.[143,255] Resident macrophages have a K_m of 0.41 mM. Casein- or sodium periodate-activated macrophages have K_m's ranging from 0.03 to 0.05 mM NADPH.[178,255,256] Monocytes actually decline in their ability to produce superoxide and hydrogen peroxide as they mature into adherent macrophages.[178,214,257,258] However, macrophages produce factors that enhance the activity of NADPH oxidase in monocytes, other macrophages, and neutrophils.[259,260] These factors include TNF, leukotriene B_4, neutrophil activating factor, C_{5a} generated by macrophage protease action on complement, and PAF (1-*o*-alkyl-2-acetyl-*Sn*-glycerol-3-phosphorylcholine).[255,261] However, alpha-interferon from macrophages can block induction of NADPH oxidase by gamma-interferon, and alpha-2-macroglobulin-protease complexes can suppress superoxide release from macrophages.[178,255] As mentioned in the neutrophil section, gamma-interferon and LPS or TNF can activate both NO synthase and NADPH oxidase activity in macrophages. Although LPS can generate direct effects, it also stimulates the production and release of TNF by macrophages and lymphocytes.[178,262,263] Unlike neutrophils, in macrophages, NADPH oxidase and NO synthase activity are reversibly turned on and off.[208] However, sustained NO synthase activity may lead to programmed cell death (apoptosis).[264] Macrophages, unlike neutrophils, may secrete thiols into extracellular medium as a defense against reactive oxygen and nitrogen oxide intermediates.[265]

2. Lymphocytes

Although it is clear from the previous discussion of macrophages that they have active NADPH oxidase, at least in certain metabolic states, activity of NADPH oxidase in lymphocytes is controversial, and contaminating macrophages are always suspected.[148,266-273] However, in a mixed population of mononuclear leukocytes dominated by lymphocytes (>90%), significant peroxidase and oxidase activities are evident.[268,274] Therefore, in functional terms, the adherent and nonadherent cells of the spleen (macrophages and lymphocytes) can be considered as an immunologic unit. Furthermore, as discussed earlier, EBV-transformed B lymphocytes do become oxidatively active.[217] Also, T lymphocyte products, like TNF and gamma-interferon, upregulate macrophage NADPH oxidase activity.[198,225,229,275,276] Again, if only chemiluminescence and peroxide production are employed to measure NADPH oxidase activity, lipoxygenase, cyclooxygenase, cytochrome P450, and NO synthase activities become difficult to distinguish from NADPH oxidase activity.[277,278]

Nonetheless, monocytic spleen cell and peripheral mononuclear cell oxidative activity can be related to the specific immune response.[279] When sheep red blood cells, rabbit red blood cells, rabbit serum, mouse red blood cells, and mouse serum were used as antigens in mice, splenic peroxidase activity peaked at two days postinoculation and returned to normal by four days.[280] When mice were immunized with bovine serum albumin, their splenic cell chemiluminescence showed significant chemiluminescence at day four postinoculation when challenged with albumin and luminol (nearly a threefold increase in chemiluminescence).[274] When human peripheral mononuclear leukocytes (>96% lymphocytes and <4% monocytes) were challenged with either a *Candida albicans* allergenic extract or influenza viral vaccine, characteristic and consistent oxidative responses occurred, as measured by chemiluminescence of luminol.[268] Influenza antigen enhanced chemiluminescence from 0.16 to 6.51 times. Immunization with influenza vaccine did not significantly affect the magnitude of chemiluminescence. Candida antigen consistently suppressed the chemiluminescent response (from −0.15 to −0.68 times control). In mice, influenza exerts some of its pathologic effects (particularly lethality) through stimulating oxidative activity that can be blocked with superoxide dismutase.[281,282] Candida antigen is known to stimulate lymphocyte proliferation and nonspecific natural killer-like cell-mediated cytotoxicity, with these T cell responses being strictly dependent on the presence of macrophages or monocytes.[283] Mouse spleen cells infected with Sendai virus have shown significantly more oxidative activity than control cells.[148] This metabolic response was generated by B and T lymphocytes independent of macrophages. Therefore, there is evidence that at least some lymphocytes contain some NADPH oxidase activity.

C. OTHER CELLS

The examination of pure NADPH oxidase systems in other cells and tissues becomes very difficult because of the close interaction of cell types that is necessary for maximum oxidative functions, the use of common substrates (NAD(P)H, thiols, oxygen, etc.) among various redox systems, and the interference of common products (superoxide, peroxides, hydroxyl radical, and NO) among the various redox systems.

1. Glial Cells and Neurons

A pure NADPH oxidase like that of leukocytes is not found in the nervous system.[241,284,285] However, NO synthase is found in glutamate and N-methyl-D-aspartate-receptor (NMDA) excitatory neurons.[60,234,286] This enzyme can leak hydrogen peroxide and, therefore, act as a weak NADPH oxidase.[70,233] Fluid-percussion injuries to the brain and ischemia-reperfusion injury stem from oxygen-radical damage as supported by evidence that superoxide, hydrogen peroxide, and hydroxyl free radical scavengers (superoxide dismutase, catalase, and dimethyl sulfoxide, respectively) minimize such injuries.[287-289] The most likely source of these reactive oxygen intermediates is not the xanthine oxidase/xanthine system found in the small intestines but rather prostaglandin H_2 synthase.[290-292] Superoxide is produced from arachidonic acid, linoleic acid or PGG_2, and NADH or NADPH as substrates.[291] In this case, the autoxidation of the NADH or NADPH is typical of that initiated by various peroxidases. It involves the single-electron reduction of organic hydroperoxide of the lipid substrate and oxidation of the NADH or NADPH to the corresponding radical. The NAD(P)-radical then interacts with oxygen nonenzymatically, yielding NAD(P)$^+$ and superoxide.[16]

The original oxidation is by an oxidized form of the PGH_2 synthase.[291] Although this enzyme usually inactivates itself upon superoxide formation within minutes, the superoxide production in nervous tissue is sustained for at least 1 hour. This effect could be attributed to the influx of granulocytes or macrophages. However, these do not increase substantially in the tissue until several hours after application of arachidonate or its cyclooxygenase or lipoxygenase products. The cerebral vasculature itself (endothelial cells) could be a source of superoxide from the xanthine dehydrogenase to xanthine oxidase conversion.[293] Also, the lipid oxidation products such as lipoxin A or leukotriene B_4 (LTB$_4$) can upregulate the NADPH oxidase activity of granulocytes.[188,294] Also, glutamate stimulation of neurons with NMDA receptors decreases uptake of cystine, leading to decreased glutathione synthesis and greater production of peroxides, superoxide, and hydroxyl radical by nonspecific metabolic processes such as the autoxidation of catecholamines.[295]

Cysteine itself is also cytotoxic for neurons.[296] The cytotoxicity is mediated through NMDA/glutamate receptor-bearing neurons.[296] Physiologic levels of bicarbonate enhance this cytotoxicity. Bicarbonate and carbonate ions facilitate the autoxidation of green hemoproteins.[63] Neurons in the cerebral cortex, hippocampus, thalamus, and striatum can be destroyed by the mechanism driven by cysteine.

Brain oxidation is further facilitated by somatotropin produced by the adenohypophysis (anterior pituitary gland). Macrophages, which may be present in the brain, are primed for enhanced superoxide production by treatment with somatotropin.[297] Glial cells may be another source of reactive oxygen intermediates or may stimulate their production in the central nervous system.[291] Microglia process antigen, have Mac-1 surface antigen markers, have nonspecific esterase activity, and secrete interleukin-1 as do macrophages.[298] Microglial oxidative activity has been demonstrated in neonatal rat microglia *in vitro*. In microglia, opsonized zymosan stimulated the greatest superoxide production with phorbol myristate acetate, 10 mM potassium chloride, platelet activating factor, and concanavalin A yielding superoxide production above that of resting cells but much less than that of cells stimulated with opsonized zymosan. Astrocytes proliferate in response to interleukin-1 but can also produce it.[299,300] When treated with gamma-interferon, astrocytes express Ia antigens and become antigen-presenting cells for Ia-dependent T cells.[299-301] Astrocytes also produce an interleukin-3-like factor that stimulates growth of peritoneal macrophages and microglia.[301] Therefore, astrocytes also have macrophage-like properties. However, their oxidative and superoxide-producing capabilities have not been demonstrated.

2. Keratinocytes and Langerhans-Dendritic Cells

Although keratinocytes and Langerhans-dendritic cells of the skin can perform macrophage-like functions, no definitive evidence has been presented for their having active NADPH oxidase.[302] However, keratinocytes can produce interleukin-1.[303] This cytokine can (1) upregulate prostaglandin synthesis; (2) enhance fibroblast proliferation; (3) enhance collagenase and phospholipase A_2 production in synovial cells; (4) act as a leukocyte chemoattractant; (5) cause degranulation of eosinophils and basophils; (6) stimulate thromboxane synthesis in macrophages and neutrophils; (7) potentiate activation of neutrophils by chemoattractants; (8) activate osteoclasts; (9) potentiate tumor necrosis factor functions; and (10) act as an autocrine, increasing keratinocyte proliferation.[303-305]

Skin fibroblasts and Langerhans-dendritic cells are antigen presenting cells like macrophages, having Class II major histocompatibility antigens, or Ia molecules, on their surfaces.[306] Therefore, these cells may indirectly activate the oxidative metabolism of leukocytes. Recently, Langerhans-dendritic cells have been shown to be oxidatively active (probably purely NO synthase).[307] Therefore, the interleukin-1 of keratinocytes may facilitate this activity. Also, keratinocytes produce macrophage-like inflammatory protein (hIP-10), which can serve as a pyrogen, be chemotactic for granulocytes, and induce them to undergo delayed oxidative bursts.[303,308,309]

3. Uterine Endometrium and Ova

The female reproductive system is unique among organ systems in having very active oxidase/peroxidase-coupled enzymes tied to a primary hormonal function.[112,113,116] Other organs such as the salivary glands, intestines, and mammary glands produce lactoperoxidase that is functional in the destruction of bacteria, but appears to lack endogenous oxidases.[6,310] Thyroid peroxidase, which iodinates thyroglobulin to form the precursor to thyroxine and triiodotyrosine, has a source of peroxide that remains undefined. The enzyme's production is controlled by thyroid stimulating hormone produced by the anterior pituitary gland. However, estrogen-induced peroxidases in vaginal, carcinogen-induced estrogen-sensitive tumors, and normal cycling uterus (endometrium) are closely linked to their primary substrate levels.[311]

The induction of uterine peroxidase by estradiol is antagonized by progesterone.[46,47,312] With physiologic levels of estrogen, uterine peroxidase begins to appear at 4 hours after induction and peaks at 20 hours.[114] Accompanying its induction is the induction of an NADH oxidase activity, which may be another activity of uterine peroxidase. The uterine peroxidase acts on the hydrogen of the catechol aromatic ring of the estradio! and artificial stilbene estrogen derivatives (diethylstilbesterol, DES).[46,47,313,314] The oxidized ring forms a free radical or quinonoid that binds to the tyrosine residues or thiol amino acid residues of proteins.[46] This reaction in essence inactivates the estrogen.[46]

Uterine peroxidase is also released into uterine and vaginal fluids where peroxide released from spermatozoa, tissues, or resident bacteria reacts with the peroxidase to generate a spermacidal or bactericidal effect.[112,113] The peroxidase was originally assumed to originate from the cyclic influx of eosinophils during the estrus cycle, in rats, but has been subsequently found localized in the endometrium of rodents and humans.[116] This cycling of peroxidase in endometrium resembles the cycling of snake skin peroxidase in the ecdysis process as described in Chapter 2. Some species of snakes also show the influx of eosinophils associated with the molting cycle.[110,315] Fetal tissue has also been shown to contain a peroxidase capable of estrogen inactivation.[316]

Whereas the sources of hydrogen peroxide for mammalian uterine and fetal peroxidases are not clear-cut, in invertebrates, such as sea urchins, there is a highly developed NADPH oxidase/ovoperoxidase couple.[317,318] In sea urchins, the fusion of gametes releases calcium divalent cations from intracellular stores.[319] This rise in intracellular calcium ion concentration results from

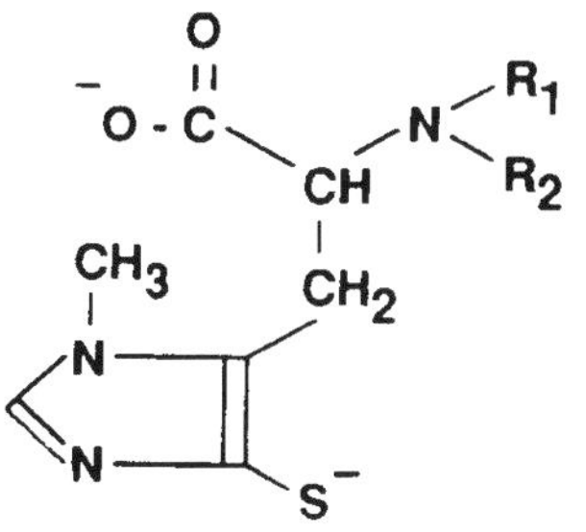

FIGURE 9. Ovothiol. A thiol from sea urchins' eggs that protects the ovum from peroxidative damage. The basic structure is 1-methyl-4-mercapto-histidine; the substituents on the alpha-amino group are: R_1 and R_2 = H for ovothiol A; R_1 = H and R_2 = CH_3 for ovothiol B; and R_1 and R_2 = CH_3 for ovothiol C.

activation of the inositol 1,4,5-triphosphate pathway, which is accompanied by an increase in diacylglycerol within 20 s of fertilization. In this process, cortical granules are excitosed from the egg. The released protein interacts with the vitelline layer in a specific orientation that is dependent on the calcium divalent cations. This fertilization envelope is then hardened by ovoperoxidase (a 70-kD protein) by oxidation of tyrosine residues into tyrosyl free radicals that recombine into dityrosine cross-links. At acidic pH's like the intracellular and immediate extracellular environment of the ovum, ovoperoxidase is inactive, but upon equilibration with seawater (a more alkaline environment), the enzyme is activated.[320] This process effectively protects the plasma membrane of the ovum from peroxidative injury.[320] The ovoperoxidase is transported from the intracellular granules to the extracellular matrix by a carrier protein that binds to ovoperoxidase with a binding constant (K_d) of 1.1×10^{-6} M.[319] The binding is dependent on calcium and magnesium divalent cations. A respiratory burst accompanies fertilization of the sea urchin egg and transport of the granule proteins.[319]

The NADPH oxidase of the ovum is cyanide insensitive (a property of b cytochromes), requires calcium and magnesium ions and ATP followed by a 60-s lag before hydrogen peroxide production (time for phosphorylation of the NADPH oxidase by protein kinase C).[319] Like the NADPH oxidase of leukocytes, the ovooxidase has membrane and soluble fractions. The soluble fraction has protein kinase C activity. Mammalian protein kinase C, but not other kinases can be substituted for the soluble fraction. Also, phorbol esters enhance the sensitivity of the ovo-NADPH oxidase to calcium ions 40-fold. To further protect the ovum from peroxidative damage, an unusual thiol, ovothiol, is produced.[320] Its general structural formula is depicted in Figure 9. These ovothiols are present in high concentrations (3–5 mM) in the sea urchin eggs. They are found in other marine invertebrates and teleost fish eggs. Glutathione easily reduces the disulfide back to the ovothiol.

Ovothiol, unlike glutathione, can react with either 2- or 1-electron oxidants. For instance, ovothiol can reduce cytochrome c (with a $k = 10^4$ M^{-1} s^{-1}), which glutathione cannot. However, ovothiols resist spontaneous air oxidation and oxidation by free ferric ions.[320]

The peroxidatively hardened envelope provides a barrier to fertilization by more than one sperm in a more elegant way than mammalian uterine peroxidase. This barrier also excludes chemical interactions with components of seawater and provides a sterile environment with surrounding cytotoxic peroxide and peroxidase.

4. Kupffer Cells and Hepatocytes

The NADPH oxidase systems of Kupffer cells and hepatocytes are very difficult to examine in isolation. Kupffer cells have NADPH oxidase (albeit at very low levels) and NO synthase capabilities as well as lipoxygenase activity.[210,321-323] The latter enzyme produces eicosanoids, leukotrienes, and organic peroxides. Hepatocytes have extracellular disulfide reductase capability, NO synthase activity, and unidentified oxidative processes stimulated by TNF and gamma-interferon.[324,325] Furthermore, these two cell types influence each others' responses to TNF and gamma-interferon, and the Kupffer cells influence the cytochrome P450 activation in hepatocytes by toxins such as carbon tetrachloride.[210,326] Furthermore, Kupffer cells also release cytokines that attract neutrophils and appear to act in concert with liver macrophages differentiated from blood monocytes that influx into the liver during inflammatory processes.[327] The latter cells are very oxidatively active. They have high NADPH oxidase activity measured by hydrogen peroxide production. They also respond as expected to TNF and gamma-interferon by producing superoxide and, subsequently, hydrogen peroxide.

The Kupffer cells do not show upregulation of NADPH oxidase with gamma-interferon and TNF, but do respond to these agents by upregulating NO synthase activity and Ia antigen expression.[328] This upregulation is enhanced by the presence of hepatocytes that also show upregulation of their own NO synthase in the absence of Kupffer cells when stimulated with these agents.[328] Hepatoma cells can produce large amounts of extracellular thiols that mask these redox activities.[265] It is apparent that although these two redox systems (NADPH oxidase and NO synthase) can be uncoupled, *in situ* and *in vivo* they probably operate at the expense and under the strong influence of each other.

III. SUPEROXIDE AND HYDROGEN PEROXIDE AS FIRST AND SECOND MESSENGERS

Finding pure superoxide or hydrogen peroxide-dependent synthetic responses of cells is very difficult because of the concomitant activation of lipid peroxidation and oxidation systems, NO synthase, thiol-producing redox systems, and phosphorylation cascades. This difficulty may be a consequence not just of imprecision in the investigators' assay methods but also of the evolutionary drive of general cellular responses to have activation and response pathways that overlap in many ways and at many levels. These fuzzy response systems provide many more options for degree and range of sensitivity to unknown stressors and stressor combinations, providing for a greater potential for survival following exotic challenges to naive cells.

As alluded to earlier, the literature is replete with examples of reactive oxygen intermediates, produced by NADPH oxidase and peroxidase and their other free radical products, causing cytotoxic effects and destroying the function and structure of biological molecules. Some of these effects are fairly specific, for example, the decrease in eye lens calcium ATPase activity by as much as 69% with only a 15-min exposure to 10^{-5} M hydrogen peroxide.[329] However, synthetic responses are harder to come by. Of these, most of the clear-cut responses are inductions of synthesis of antioxidants, antioxidative enzymes, and DNA repair enzymes.

A. OXIDATIVE STRESS GENES

Free radical and oxidative damage to DNA induces the classic SOS response leading to excision and recombination repair of segments of damaged DNA nucleotide residues.[240,330,331] These DNA repair processes involve endonucleases (like III and IV), lyases and ligases, glycosylyases, and phosphatases.[332] The induction of antioxidant enzymes by hydrogen peroxide in *Escherichia coli* involves damage to a protein, OxyR, which when oxidized binds to different DNA residues than the reduced form, derepressing expression of an operon which is transcribed and translated into catalase (hydroperoxidase (oxyR) I katG gene), glutathione peroxidase (gorA gene), and components of alkyl hydroperoxidase (ahp gene).[219,240] The components of the ahp gene complex include glutathione reductase, C22 (ahp C gene) and F22 (ahp F gene). The alkyl hydroperoxidases of bacteria are functionally and structurally similar to the mammalian enzyme glutathione peroxidase but lack selenium. In total, oxidation of oxyR is responsible for induction of about 30 proteins detectable by two-dimensional electrophoresis, 12 of which are seen maximally within 10 min. The oxidative activation of oxyR is reversible with dithiothreitol reduction.

Another oxidative stress operon triggered by paraquat and other superoxide-generating reductants is soxR (nine proteins respond to its activation).[240] This regulatory protein upregulates manganese superoxide dismutase (sodA gene) and endonuclease IV (nfo gene). It also controls expression of glucose-6-phosphate dehydrogenase (which supplies NADPH to reductases and oxidases), S6 (the small ribosomal subunit protein, or protein synthesis protein) and OmpF (the membrane permeability factor that regulates absorption of antibiotics). The effects of this control by soxR is upregulation of the soxR operon in response to redox cycling agents (like quinonoids), decreased concentrations of NADPH, and the presence of antibiotics. The soxR locus also contains another controlling gene soxS. There is an 85-bp intervening space between the head-to-head positioning of these genes. The protein products of soxR and soxS jointly control the operon.

Another operon triggered by hydrogen peroxide is katF.[219] This gene expression is also sensitive to near-ultraviolet light radiation (300- to 400-nm wavelengths). Besides catalase (hydroperoxidase II; katE gene), Bo1A, a morphogenesis protein, acid phosphatase, and endonuclease III (xth gene) are also regulated by the katF gene product. In the stationary phase of growth, the katF protein and the proteins of the genes it controls are maximally synthesized. This response is in a sense anticipating oxidative injury in the saturated culture.

Besides oxyR and katF, other oxidative stress genes include fur and arcA, which also regulate the manganese superoxide dismutase (MnSOD).[219] Bacteria also have an iron superoxide dismutase, but it is constitutive (sodB gene). The MnSOD is found strictly in the inner membrane of mitochondria of eukaryotes. In yeast, the expression of this gene is also controlled by superoxide or another reactive oxygen intermediate.

B. ANTIOXIDANT ENZYMES

The major cytosolic superoxide dismutase of eukaryotes is copper/zinc superoxide dismutase (CuZnSOD).[145,333] Although predominantly in the cytoplasm of liver cells, endothelial cells, and smooth muscle cells, a minor percentage of CuZnSOD is found in the nuclei of cells. In human fibroblasts, hepatoma cells, and yeast cells, the enzyme is found predominantly in peroxisomes. It is also found in the granule-rich fraction from canine pancreatic beta cells. It is also found in the tissues of many diverse species, including maise, the fruit fly Drosophila, rat lung, red blood cells, wheat, kidney beans, tomatoes, and spinach. An extracellular CuZnSOD (at least three isozymes) is found in plasma, lymph, and synovial fluid of man and other mammals.[333]

Hyperoxia in animals induces catalase, glutathione peroxidase, and superoxide dismutase (CuZnSOD) synthesis.[333,334] Bleomycin, a superoxide generator, strongly elevates the total and specific activity of SOD in the lungs of hamsters. Endotoxin induces MnSOD in pulmonary artery endothelial cells, which are the target of most lung hyperoxia damage. Chronic hyperoxia causes a gradual increase in glutathione peroxidase, glutathione reductase, and glucose-6-phosphate dehydrogenase (source of the NADPH for the glutathione reductase).

The importance of antioxidant enzymes in human pathology is further emphasized in conditions like Down's syndrome and Lou Gehrig's disease (amyotrophic lateral sclerosis; ALS).[335] In both cases, chromosome 21 is involved. In Down's, trisomy of chromosome 21 occurs, and there is an overabundance of CuZnSOD. Perhaps, this defect leads to overproduction of hydrogen peroxide by the SOD and, subsequently, peroxidative damage. In inherited ALS, there is a defect in the CuZnSOD gene, perhaps leading to superoxide accumulation and subsequent damage.

There is a neurogenic peroxidase in the brainstem that is cytotoxic to various cells when activated with hydrogen peroxide or NAD(P)H and manganese ions and/or kainic acid (as an oxidant).[336] Its cytotoxicity is strongly inhibited by dopamine. This peroxidase may play a role in Parkinson's disease and may also be responsible for melanin formation in the substantia nigra.[336] There is strong evidence that peroxidase is an important catalyst for the formation of melanin in humans and other mammals.[337,338] The source of the hydrogen peroxide for this process in the nervous system is likely to be monoamine oxidase.[295]

Since copper ions can induce the autoxidation of ascorbate, catecholamines, and other biocompounds, the regulation of CuZnSOD synthesis is coupled to metallothionein induction (a transition or heavy metal chelator in cells).[334] Proteins called ACE1 are involved with the coinduction with copper ions of metallothionein in yeast cells by interaction with the CUP1 gene for metallothionein.[339] In yeast lacking a functional ACE1 gene, the upregulation of CuZnSOD with addition of copper ions does not occur. This response suggests that the CUP1 and SOD1 (gene for CuZnSOD) are coregulated by ACE1. Anaerobic conditions also lead to less uptake of copper ions by yeast cells and production of nonfunctional CuZnSOD apoenzyme. The ACE1 binding site in the genome is upstream from the promoter of SOD1. Therefore, free copper ions that might be oxidatively active are prevented from accumulating by this coordinated gene expression.

C. HEME OXYGENASE

Further evidence for the interaction of oxidative stress and transition metal ion metabolism is found in the process of heme oxygenase induction.[51,52] This unlikely stress protein (SP32) is also associated with cellular responses to long-wavelength UVB (290 to 320 nm) and UVA (320 to 380 nm), short-wavelength UVC (<290 nm) and near-visible UV (405 nm).[52] UVA and UVB irradiation can oxidatively activate extranuclear targets such as flavins (which may generate superoxide and peroxides), tryptophan (which may generate hydrogen peroxide), and NAD(P)H (which can generate superoxide).[128,341,342] Irradiation with UVA or UVB of porphyrins can generate singlet oxygen, which can attack unsaturated fatty acids to form lipid peroxides.[27] The superoxide and hydrogen peroxide can interact through a transition metal ion in the Fenton reaction, generating highly reactive, short-lived hydroxyl free radicals.[238]

Heme oxygenase was recognized as a stress protein (SP32) because it responds to stresses that appear only distantly related to heme breakdown.[52] Besides the oxidant hydrogen peroxide, sulfhydryl reactive agents like sodium arsenite and other heavy metals and the tumor promoter TPA also induce SP32.

Iron chelators such as desferrioxamine and phenanthroline suppress mRNA transcription and translation of SP32 induced by UVA irradiation or hydrogen peroxide treatment of human fibroblasts. This induction was maximal at 2 hours and dropped to baseline at 6 hours.[52]

This induction of heme oxygenase has also been seen in human primary and transformed keratinocytes, lung and colon fibroblasts, HeLa and lymphoblastoid cells, and in monkey, hamster, rat, mouse, and opossum cells.[52] Therefore, the response is considerably generalized among cell types and species. However, the response was lowest in epidermal keratinocytes and highest in cultured human lymphoblastoid cells. These results parallel the increased translation of ferritin for binding up intracellular iron released by the heme oxygenase and downregulation of transferrin receptor translation by mRNA degradation.[49,343]

Heme oxygenase is in itself not a heme protein but binds heme and mimics cytochrome b heme autoxidation without release of superoxide or peroxide.[51] The microsomal heme oxygenase system utilizes a NADPH-cytochrome c reductase (fpT) to provide the two single-electron transfers necessary to the heme autoxidation. When cytochrome b_5-NADH-dependent reductase was substituted for the fpT, it could reduce the ferric heme but not drive the autoxidation to biliverdin and iron ion release.

The presence of the heme and iron processing proteins and their reactions along with the cytochrome b subunit of the NADPH oxidase in many cell lines suggests a general role for a cytochrome b-linked sensor of oxidative injury. But, also, the cytochrome b sensitivity to heme loss due to heat, pH, and perhaps, other nonspecific denaturing processes suggests that a nonoxidative mechanism may trigger a similar pathway to the oxidative ones. However, heme oxygenase is not induced by heat alone.[52] When the apoprotein of hemoglobin (globin without the heme) was injected into cells, it initiated translation of heat shock protein (HSP70).[220,344] Therefore, both the heme and globulin components of a heme protein sensor could independently activate different stress proteins. Furthermore, reduced glutathione (GSH) inhibits heme oxygenase induction, and its depletion increases heme oxygenase induction. Therefore, SP32 induction can be blocked by high thiol levels and should be heat sensitive (from heme release from denatured heme protein) when cells are thiol (reduced glutathione) depleted.

Are there other examples of direct interaction between diverse stimuli to yield cross-protection? Yes, one such example is seen in the retina.[345] When rat retinas were treated by heating to at least 41°C for at least 15 min, HSPs 110, 74, and 64 increased above baseline and the retinas were protected from light-induced photoreceptor degeneration. The rats were exposed to 2690 lx of light for 24 hours in a 30°C chamber. Protection was demonstrated when the rats' retinas were exposed to the light 10 to 20 hours after hyperthermia with a completely protective response at 18 hours and a return to nearly baseline response at 50 hours.

Other conditions that yield heat shock proteins in the central nervous system are treatments with D-lysergic acid diethylamide (LSD) or sodium arsenite, or ischemia, surgical cutting, and concussive injury.[220,300] Furthermore, glial cells produce heat shock proteins at the site of stress or injury and transfer them directly to neuronal axons. Therefore, the stress response shows much overlap between oxidative and nonoxidative injury.

Other direct effects of oxidants on mammalian cells have been observed. These include induction of c-fos expression and DNA synthesis in BalB/3T3 fibroblasts. The synthetic response of DNA may be through ribonucleotide reductase, the key rate-limiting enzyme in DNA synthesis.[346] The deoxynucleotide production by this enzyme is dependent on the formation of a stable tyrosyl radical on the enzyme that disappears in cells held in a higher reducing or anaerobic state, but reappears in the presence of oxygen.[346]

D. OXIDATIVE STRESS AND THE HUMAN IMMUNODEFICIENCY VIRUS

Another direct effect of activation of an extranuclear receptor or sensor in the plasma membranes of cells is seen in the growth of mammalian cells and expression of retrovirus. The upregulation of human immunodeficiency virus long terminal repeat sequence (HIV-LTR) expression labeled with the chloramphenicol acetyltransferase reporter gene has been reported to be effected by UVC (200 to 280 nm wavelength radiation).[347] The level of upregulation, measured by chloramphenicol acetyltransferase (CAT) activity, was 50- to 150-fold baseline. Direct sunlight activated this expression 12-fold. UVB (280–315 nm) also upregulated HIV. Transgenic mice carrying HIV with a beta-galactosidase or firefly luciferase reporter gene showed upregulation of HIV after exposure to either UVB or UVA (315–400 nm) light combined with the photosensitizer psoralen.[348] Sunlight, which contains both UVA and UVB, worked without photosensitization but required 7 hours as opposed to 2 hours with psoralen added. The Langerhans cells of the skin, which mature into antigen-presenting dendritic cells, which can efficiently infect CD4+ T lymphocytes with HIV, could potentially harbor virus that would be activated by UV light. Again, the conditions for HIV activation by UV radiation strongly resemble those necessary for the induction of heme oxygenase. Either this is a happenstance of multiple irrelevant gene activation or a link to a particular sensor (redox or heme related).

Evidence for the existence of a plasma membrane extranuclear sensor has been provided by data on the UV activation of the broadly acting transcription factor, NF-kappaB.[349] The UV activation of binding of this factor to DNA has been shown without nuclear material being present.[349] Furthermore, the same factor is upregulated by exposure of cells to nitric oxide and other oxidants. Its activation is inhibited in part by antioxidants at the level of the GTP-binding protein Ha-Ras.

NF-kappaB is the endogenous transcription factor responsible for activating HIV expression.[349] Therefore, oxidants like NO or UV radiation, which activate NF-kappaB, could activate HIV expression *in vivo*. Ionizing radiation (^{60}Co gamma irradiation at 700 for 10 min) activates NO synthase.[350] Therefore, oxidations initiated by ionizing irradiation are expected to upregulate HIV expression. Gamma-interferon, which activates oxidative activity of macrophages, should also upregulate HIV expression. LPS has been shown to upregulate latent Abelson leukemia virus expression in RAW 264.7 mouse monocyte/macrophage cells.[351] Endotoxin activates both oxidative and reductive pathways in RAW 264.7 cells.[218] These events could activate Abelson leukemia virus expression through a NF-kappaB pathway like that of HIV.

Oxidative processes and UV radiation can also act indirectly on the upregulation of cell growth and latent retrovirus expression. Exposure to 60 min of 20 J/m^2 of 254-nm UV enhanced NF-kappaB binding activity fivefold by 60 min, with the activity continuing to rise to tenfold by 8 hours.[349] Phorbol ester-treated cells had the same NF-kappaB activity response. When HIV-I-negative HeLa cells were irradiated with 45 J/m^2 UVC and cocultured with HIV-I-positive HeLa cells, there was a three- to sixfold increase in HIV-LTR upregulation.[352] Conditioned medium from irradiated cells also transferred this stimulatory activity. These effects were not associated with cell proliferation. CD+4 T cells (Molt-4) and Jurkat cells also responded to these conditions by upregulating HIV-1. The UV-induced or activated factor has been detected in several other cell types, including normal human primary skin fibroblasts, skin fibroblasts from xeroderma pigmentosum patients, and the human hepatoma cell line HepG2.

Interleukin-2 (IL-2) production by CD+4 T lymphocytes is also controlled by NF-kappaB.[353] Therefore, agents that stimulate NF-kappaB activation can act as adjuvants to immune stimulants. However, the homodimer complex of the NF-kappaB subunit p50 is inhibitory to IL-2 induction.[354] An unknown protein must be activated to dislodge the p50-p50 complex and replace it with the active NF-kappaB p50-p65 active inducer. NF-kappaB activation in transformed lymphocyte cell lines such as EL4 cells can be induced by PMA; in transformed lymphocyte cell lines such as Jurkat cells it can be induced by lectins. Both of these are partial immunologic signals.[354] Nontransformed T lymphocytes become anergic with such partial signals and require full antigen stimulation.[354] This full stimulation includes interaction with antigen plus antigen-presenting cells for displacement of p50-p50 in the nucleus by p50-p65 NF-kappaB. These negative and positive regulatory complexes of NF-kappaB are found in lymph node T cells and thymocytes.

Stressed RAW 264.7 mouse cells demonstrated the production of factors in their conditioned medium supernatants that increase viability and proliferation of other RAW 264.7 cells and the survival of cytotoxic mouse lymphocytes (CTLL-2) activated by IL-2 *in vitro*.[218] The RAW cells were stressed by cultivation at high cell-population density, treatment with LPS, treatment with microwave radiation or hypo- or hyperthermia. This conditioned medium or stressors upregulated endogenous latent Abelson leukemia virus in the RAW 264.7 cells or HIV-LTR-CAT gene expression in HeLa cells.[351] The release of factors was associated with oxidative stress followed by enhanced extracellular thiol production. Two factors (one associated with cell viability and the other with viral expression) were separable from conditioned medium by ion exchange chromatography.[355]

The support afforded by the feeder cell activity of RAW 264.7 and Novikoff rat hepatoma cells to CTLL-2 cells was associated with extracellular thiol production.[265] Another leukocytic cell line, HL-60 promyelocytic human leukemia cells, produces a 13-kD autocrine under high cell-population density stress that also stimulates cell growth.[356] Primary cell cultures of Kaposi's sarcoma lesions of AIDS patients constitutively produced and released growth factors—basic fibroblast growth factor, a potent angiogenic factor, and interleukin-1 (IL-1)—that act as autocrines and paracrines (induce sarcoma-like lesions in surrounding tissue).[357]

The interactions among cytokines, HIV, and other viruses and cells (both virally infected and not infected) are two way. The U1 clone of U937 human promonocytic cell line chronically infected with HIV-1 was upregulated by granulocyte/macrophage colony stimulating factor, but not by IL-1, gamma-interferon or tumor necrosis factor alpha.[285,358] IL-1-beta was seen in this clone but not in the parental line. These responses to cytokines may have been unique to the stage of differentiation of this monocytic cell line or reflect virally altered responses. Viruses may mimic cytokine, lymphocyte, and/or macrophage stimuli. HIV-1 is a strong human B lymphocyte polyclonal activator like LPS.[359-361] Elevated serum levels of immunoglobulin G (IgG) and immunoglobulin A (IgA), immune complexes, and peripheral blood lymphocytes spontaneously secreting immunoglobulins are seen in AIDS patients. Normal human B lymphocytes incubated for 1 hour with HIV showed marked proliferation and differentiation, which peaked by 4 days and did not depend on accessory cells (antigen-presenting cells; APC). A dramatic illustration of this cross-reactivity of stimulants was seen in macrophages stimulated with gamma-interferon or cytomegalovirus.[362] The latter stimulates macrophages even when the virus is inactivated. Class I-restricted CD+8 T lymphocytes raised against the 65-kD heat shock protein of *Mycobacterium bovis* recognized stressed bone-marrow-derived mouse macrophages, resulting in their lysis (an autoimmune response).[363] Human cytomegalovirus induces protooncogene c-fos, c-jun, and c-myc expression in infected human cells without viral protein synthesis.[362] This virus also increases cellular inositol trisphosphate and 1,2-diacylglycerol production, calcium ion influx, increased cytosolic calcium activity and increased cellular levels of cyclic adenosine 3′,5′ monophosphate. Thus the cellular response systems to cytokines, bacterial products and viruses have overlapping pathways and are at least in part involved with redox pathways.

E. CYTOKINE INTERACTIONS

Cytokines such as the monokine TNF alpha and interleukin-1 can directly induce effects that appear related to redox activity but in fact are not. For instance, TNF increases transcription of MnSOD mRNA in rat lung while having no such effect on CuZnSOD.[364] Similarly, IL-1 increases transcription and translation of MnSOD and gradually increases expression of CuZnSOD in lung fibroblasts.[364,365] TNF and IL-1 treatment increases survival of rats exposed to hyperoxia and protects murine hematopoiesis from ionizing radiation damage when administered before irradiation.[366-368] Glutathione reductase, glutathione peroxidase and catalase are also induced but, unlike MnSOD, they remain in elevated quantities for a long time after the inducer is removed.[364,365] These effects are comparable to a response to a "phantom" oxidative stress.

Ras oncogene activation by missense mutations leads to substantial significant intrinsic ionizing radiation resistance in transformed and nontransformed cell lines, whether the altered oncogenes are introduced by mutation, infection, or transfection.[369] Ras effects nerve growth factor and other neurohormone and IL-1 interactions and feedback regulation. This effect on ras and its functions is again reminiscent of UV or NO effects on NF-kappaB activation.

The cross-reactivity and multiple functions of IL-1 appear on the surface to be evidence for physiologic control of the immune system through feedback onto the endocrine system. Glucorticoid hormones inhibit the effects of IL-1 as an endogenous pyrogen, stimulator of acute-phase protein production by hepatocytes, augmentator of superoxide production by granulocytes, and modulator of fibroblast growth and of collagenase and prostaglandin production.[365,370] IL-1, but not TNF, IL-2, or gamma-interferon, stimulates the pituitary gland to secrete increased levels of adrenocorticotropic hormone, which, in turn stimulates the increased production and secretion of glucocorticosteroids by the adrenal glands in rats and mice.[371] Dexamethasone, an artificial glucocorticosteroid, increases the content of superoxide dismutase and catalase in rat lung, which is a result of stimulated transcription rate of mRNA for these two enzymes as opposed to affecting mRNA stability. IL-1 activates ACTH release at the level of the brain, stimulating the hypothalamic neurons to release corticotropin-releasing factor (CRF).[372,373] The IL-1 has no direct effect on the pituitary or adrenal glands. Other hypothalamic hormones are unaffected by IL-1 stimulation of the hypothalamus.[372-374] However, recombinant human IL-1-beta directly stimulated rat pituitary cells *in vitro* to secrete ACTH, luteinizing hormone, growth hormone, and thyroid-stimulating hormone.[371] The IL-1-beta levels necessary for this stimulation of secretion were 10^{-9} to 10^{-12} M (serum levels). Prolactin secretion (an immunostimulatory hormone) was inhibited by these concentrations of IL-1-beta (perhaps through ACTH feedback).[371] Other papers contradict this direct effect on the pituitary of IL-1.[372,373]

This simplistic feedback mechanism between the immune system and the nervous system was complicated by the discovery of endogenous IL-1 production in the brain.[374] Glial cells were first found to be a source of IL-1, followed by neurons. Using immunochemical techniques (labeled antibody against recombinant human IL-1-beta), IL-1-beta neurofibers (but few if any IL-1-alpha) were immunostained in the hypothalamus of human brain. The densest accumulation of stained fibers was in the periventricular regions that participate in pituitary control.[372,374] Immunostained fibers were found in the periventricular and arcuate nuclei and in the parvicellular part of the paraventricular nucleus. The cell bodies of neurons that produce corticotropin-releasing factor are in the paraventricular nucleus and were heavily stained with IL-1-beta immunoreactive antibody. Stained fibers extended into the infundibulum and the region of the median eminence that contains the hypophyseal portal vessels. IL-1 immunostaining fibers were seen in the magnocellular part of the paraventricular nucleus in cell bodies of neurons that secrete oxytocin and arginine vasopressin from the posterior pituitary gland.

Il-1-beta immunoreactive fibers were also found in areas of the brain responsible for central autonomic control, including autonomic portions of the paraventricular nucleus, the dorsomedial nucleus of the hypothalamus, the lateral hypothalamic area, the subfornical organ, the bed nucleus of the stria terminalis, the substantia innominata, the ventromedial nucleus of the hypothalamus, the posterior hypothalamic area, and the paraventricular nucleus of the thalamus.[372,374] Areas responsible for innervation in the febrile response and the acute-phase reaction contain IL-1-beta immunostimulatory neurons. For the febrile response centers, this staining included the periventricular preoptic nucleus, which contains thyrotropin-releasing hormone neurons. Portions of the paraventricular and supraoptic nuclei that contain vasopressin-producing neurons were also stained. Also, there are hypothalamic cell bodies associated with central cardiovascular regulation that are IL-1-beta immunoreactive.

Neuronal control for liver secretion of c-reactive protein, serum amyloid A, and ceruloplasmin is located in the ventromedial nucleus and lateral hypothalamus. Both these area are innervated by IL-1-beta neuronal fibers. The lateral hypothalamus area also innervates the entire cerebral cortex and may be responsible for the IL-1 effects of drowsiness accompanied by electroencephalographic synchronization. Most of these areas discussed are distal from the organum vasculosum of the lamina terminalis of the hypothalamus. This area is believed to be the portal of entry of circulating exogenous IL-1 into the brain because it has no blood-brain barrier (a circumventricular organ located at the anteroventral tip of the third ventricle).

This observation led to the proposal that exogenous IL-1 stimulates production of endogenous IL-1 in the brain (a positive-feedback system). In animals, injection of morphine into the periaqueductal gray matter results in rapid suppression of natural killer cell activity.[375] This effect can be blocked by the opiate antagonist naltrexone and is probably mediated through the neurohormone mechanisms discussed above.[375]

The cross-reactivity between the immune system and the nervous system may run even deeper than the aforementioned discussion illustrates.[376,377] The highly conserved neurohormone alpha-melanocyte-stimulating hormone (MSH a neuropeptide) is antipyretic, inhibits histamine-induced increases in vascular permeability and acts as an anti-inflammatory.[378] The latter property was demonstrated by its inhibition of ear swelling in mice injected with the irritatant picryl chloride (the inhibition was equal or better than with prednisolone). Circulating levels of alpha-MSH are increased by intravenous injection of IL-1 and endotoxin.[365,378] Another example of the cross-reactivity is found in the fact that polyclonal activation of B cells by HIV is mimicked by a neuropeptide, neuroleukin (NLK), which is secreted by activated T lymphocytes as well as nervous tissue.[379] Neuroleukin has sequence homology with sections of the immunosuppressive HIV protein gp120; this protein also competitively inhibits neuron growth in culture supported by NLK.[379] Therefore, AIDS dementia and the stimulation of the Thy-2 (humoral-generating helper T cells) portion of the immune system may be related to this cross-reactivity.[380,381]

Another two systems that overlap with each other and with the immune system through redox-regulating cytokines are the female reproductive system and the skeletal system. These interactions extend to other bystander organ systems. For instance, superoxide dismutase and catalase activities are elevated in Syrian hamster kidneys in response to estrogen treatment, indicating the hormone is mimicking a redox signal or generating one that induces the antioxidant enzymes.[313,314,382] As stated earlier, estradiol induces production in endometrium of the oxidase/peroxidase system that leads to the hormone's covalent linkage to cellular protein and its inactivation. Furthermore, the fundamental chemical structure of estradiol, a complex polycyclic aromatic phenol, makes it an ideal peroxidase co-factor. It stimulates the autoxidation and chemiluminescence involved in the oxidation of NAD(P)H by horseradish peroxidase.[110] However, again, like the action of IL-1, the action of this hormone goes beyond a simplistic induction of a limited number of genes in the endometrium. Three diseases, with apparently unlikely relatedness, illustrate the pluripotential interaction of estrogen with the skeletal, immune, and reproductive systems.

The first of these diseases to be addressed is osteoporosis, the major cause of postmenopausal disabling bone fractures in women.[383] It is due to the deficiency of 17-beta-estradiol. Osteoclasts (derived from monocytic progenitor cells), the primary effector cells in osteoporosis, are activated by cytokines released from osteoblasts activated by parathormone (PTH), IL-1, and TNF. Estradiol and progesterone inhibit this activation by inhibiting production of the osteoblast cytokines. Peripheral blood monocytes and macrophages also show enhanced production of IL-1, TNF, and granulocyte-macrophage-colony-stimulating factor (GM-CSF) when estrogen levels are low. These monokines not only activate osteoblasts, but also cause differentiation of osteoclast precursors directly. Estradiol and progesterone also directly decrease the production of these cytokines by monocytes and macrophages. The osteoblasts and/or stromal cells when activated by PTH or the monokines also produce macrophage-colony stimulating factor (M-CSF), GM-CSF, and interleukin-6. Testosterone and progesterone inhibit this cytokine production to a lesser extent than estrogens.

Osteitis deformans (Paget's disease in humans) is an incidental bone disease of people in the Western Hemisphere beginning in their forties and of snakes.[384] It has both the appearance of an inflammatory process and a cyclic endocrine disease. The lesions are very localized in bone with episodic osteoclasis and bone collapse followed by excessive deforming dense bone formation. This disease has been associated with excessive IL-6 production in humans and chronic unresolved Gram-negative bacterial infections in snakes.[364,384,385] Again, it appears that cytokine and hormonal effects in this disease have the same outcome as an inflammatory process. Associated with osteitis deformans is spontaneous fracture and metastatic normal bone marrow.

Endometriosis, the third disease to be examined in this set, is characterized by metastasis of untransformed endometrium outside of the uterus.[386,387] Women and female rhesus monkeys both develop spontaneous endometriosis. Breast cancer and endometriosis development have been associated with polycyclic hydrocarbons such as 2,3,7,8-tetrachlorodibenzo-*p*-dioxin, which may have some estrogen-mimicking properties.[313,314,382,386] When peripheral blood mononuclear cells (PBMC) from these monkeys with endometriosis were examined, investigators found that the PBMC from monkeys with endometriosis produced significant levels of TNF upon endotoxin stimulation (20–97 pg/ml). Control monkeys' PBMC, from those without endometriosis, did not produce TNF-alpha when stimulated with endotoxin. In the diseased monkeys, the TNF was not produced spontaneously. Both disease-free animals and those with endometriosis produced IL-6 upon LPS stimulation in these studies. Monkeys with severe endometriosis produced less IL-6 upon stimulation than did controls (101 vs. 376 pg/ml). The low levels of IL-6 may have led to less terminal differentiation of endometrial cells, allowing for their proliferation and maintenance. TNF may have acted as a stimulator of fibroblastic (stromal) secretion of growth factors that has escaped estrogen suppression. Finally, the basis for paradoxical and multiple effects of these redox modulators needs to be addressed.

The simplest explanation for these differential and often contradictory responses to cytokines is that they exist as various isoforms with various activities. An example of this diversity involves IL-1. Activated mononuclear cells are implicated in the development of autoimmune beta cell (in the Islets of Langerhans of the pancreas) destruction in insulin-dependent (type I) diabetes mellitus. This cytotoxicity is mediated through IL-1.[388] However, whereas the isoform pI7 of IL-1 inhibits insulin production, the isoforms pI6 and pI5 of IL-1 show no effect.[388]

Another simple modulation of a redox effector molecule is its interconversion. For instance, eosinophils, when oxidatively activated, or eosinophil peroxidase plus hydrogen peroxide and halides convert the slow-reacting substance and vasoactive leukotriene LTC$_4$ to the leukotriene LTB$_4$, which stimulates the oxidative burst in granulocytes and activates lymphocytes.[188,389] An example of oxidative modulation involving protein is found in C-reactive protein. The oxidative denaturation and activation of proteolytic activity by such modification is well documented. However, specific oxidative enhancement is less evident. C-Reactive protein is a major acute-phase protein in human serum.[390] When treated with a ferrous/cupric ascorbate system (which generates reactive oxygen intermediates), C-reactive protein converts into a form that enhances (allows the initiating factors to work more efficiently at lower concentrations) the aggregation of and serotonin release from human platelets by platelet activating factor (PAF: 9.1×10^{-9} M), thrombin (0.1 U/ml), ADP (0.5 μM), epinephrine (20 ng/ml), the calcium ionophore A23187 (10 μM), or arachidonic acid (0.1 mM). The oxidized human C-reactive protein did not effect collagen-induced platelet activation of rabbit platelets (therefore, species specific).[390]

Another example of mediator oxidative modulation is found in the immunosuppressive activity of polyamines, which are normally associated with rapid cell division and neoplasms.[391] Spermine oxidase from fetal calf serum was shown to convert spermine and spermidine into oxidation products that suppressed LPS-induced mitogenesis of CBA/J mouse spleen cells and the murine mixed lymphocyte reaction. Also, a spermine oxidase from mouse amniotic fluid was found to produce noncytotoxic immunosuppressive agents from cadaverine, spermidine, and spermine.

Another possible regulatory level for a cytokine-controlled redox response is at the effector enzyme level. The spin trap 5,5-dimethyl-1-pyrroline-*N*-oxide leads to a dose-dependent depolarization of resting human neutrophils and prevents depolarization by membrane-active agents like *N*-formyl-methionyl-leucyl-phenylalanine (FMLP) or concanavalin A without affecting intracellular calcium ion release or extracellular calcium influx accompanying the stimulus.[392] This spin trap brings about conditions for examining direct field effects (changes in membrane potential) on the activity of the NADPH oxidase. Oxidase-activated phenols and peroxidase-activated methoxyphenols also interfere with human neutrophil NADPH oxidase assembly, or activation.[200,201] These effects are dependent on oxidative processing of the respective phenol. The quinones formed from the phenol evidently bind to critical thiols in the NADPH oxidase complex.[53] 3-Amino-L-tyrosine evidently inhibits the oxidative burst of macrophages by conjugation of its quinonoid oxidation product with critical enzyme nucleophiles (bases such as thiolate or amines) in the oxidase and peroxidases.[72,117] A similar self-oxidative inactivation is seen in prostaglandin H synthase, which self-acetylates when treated with aspirin.[393] This inactivation by acetylation is dependent on the presence of the prosthetic heme group of the enzyme.

Cytokines or factors that turn on one activation pathway may simultaneously turn on or off another pathway leading to altered sensitivity to other inducing agents or revealing previously dominated alternate (minor) pathways. Several examples of these kind of effects can be given. First, activation of the cytotoxicty of the mouse macrophage cell line RAW 264.7 toward the target lymphoid cell line P815 can be measured by the release of radiolabel.[394] When the RAW cells were treated with conditioned medium from murine spleen cells of C3H/HeN mice treated with Con A (2 µg/ml) and suboptimal concentrations of LPS (1 ng/ml) for 72 hours, they demonstrated release of the RI regulating subunit (45 kD) from cAMP-dependent kinase I (cAMP-PKI). Treatment with LPS alone at high concentrations (10 ng/ml or 100 µg/ml) resulted in the same effect. The lymphokine in the conditioned medium only prepared the RAW cells for the response, but did not result in the cAMP-dependent kinase activation or cytotoxic activation until LPS was added as a triggering agent.[394] Forty-eight hours after removal of these stimulants, the RAW cells lost their target-cell cytotoxicity and showed a re-expression of the RI 45 kD in the cytosol.

Another example of multiple effects of cytokines is found with prostaglandin E_2 (PGE$_2$; 1 µM). PGE$_2$ has been shown to activate cAMP-PKI activity by increasing cAMP levels (7.5-fold) within 30 s in the mouse macrophage-like cell line P388D$_1$.[185] These cAMP levels dropped rapidly to 100 pmol/10^7 cells, stayed at this level for 20 min, and gradually increased to fivefold above the control at 200 pmol/10^7 cells. Neither PGF-alpha nor PGD$_2$ activated PKI. In RAW 264.7 cells, LPS (100 pg/ml) or lipid A stimulated protein tyrosine kinase with protein phosphorylation being detectable at 4–5 min, peaking at 15 min and declining after 30–60 min.[186] This stimulation was maximal at a dose of 10 ng/ml LPS. It was related to arachidonic acid metabolite release, which in turn could have led to PGE$_2$ activation of cAMP-PKI.[394] The latter event would have been accomplished through PGE$_2$ receptor-mediated activation of adenyl cyclase and the subsequent production of cAMP.[198] However, arachidonic acid release is concomitant with and supportive of macrophage activation, but not essential to it.[181,194] As we reported earlier, arachidonic acid and perhaps protein phosphorylation can interfere with the binding of the GAP protein associated with the NADPH oxidase G-protein, which in turn triggers adenyl cyclase, which produces cAMP. This interference effectively allows for G protein activation without also activating the GTPase activity, which turns the G-protein off. Con A, wheat germ agglutinin, and thrombin, which stimulate arachidonic acid release, do not upregulate hydrogen peroxide production by mouse peritoneal macrophages.[194] Also, the type of cell stimulated may determine the dominant pathway triggered by agents like LPS.

Individual cell types show different pathway dominance with the same stimulant.[198] The P388D$_1$ cells do not require a tyrosine kinase for LPS-mediated activation (as supported by inhibitor studies).[196] Also, tyrosine phosphorylation does not appear to mediate LPS activation in B cells.[395] As alluded to earlier, LPS can activate the phospholipase C/protein kinase C pathway.[182,191]

Multiple pathway stimulation by cytokines that modulate redox metabolism is further complicated by feedback controls. IL-1 produced by macrophages upon LPS stimulation induces PGE_2 production through cyclooxygenase.[370,396] However, addition of PGE_2 or PGI_2 to macrophages decreases the yield of IL-1, demonstrating a classic hormonal feedback loop that ultimately affects oxidative metabolism (leukotriene and lipoxygenase do not participate in this feedback).[396] Fibroblast growth factor (FGF) production by CBA/J mouse peritoneal macrophages has also been shown to decrease in LPS-induced cells (1 µM of PGE_2 decreased FGF by fourfold).[198] TNF treatment has also led to increased PGE_2 production, which in turn, increased cAMP levels and metabolic activation of mouse peritoneal macrophages.[198] Indomethacin, a cyclooxygenase inhibitor, has been shown to inhibit the effects of TNF while PGE_2 has displayed enhancement of TNF effects and has reversed indomethacin inhibition. Cholera toxin, an adenyl cyclase stimulator, and dibutyl-cAMP also enhance TNF effects on metabolism.[198]

The anti-inflammatory and neutrophil-inactivating effects of adenosine released from ischemic cells undergoing ATP depletion also support the central role of cAMP-dependent kinase in these intersecting pathways.[187,206] Other co-factors, produced in conditioned medium by peripheral human lymphocytes (T cells) stimulated with Con A but which are not gamma-interferon, IL-1, or TNF, stimulate NADPH oxidase activity.[276] Human monocyte-like transformed cells, HL-60 and U937, also release these co-factors into their supernatants when stimulated with Con A. One such lymphokine (54 kD M_r) requires 48 to 72 hours to upregulate the NADPH oxidase response.

Other examples of competing, interacting, or interfering pathways are found in the interactions of phorbol esters, chemotactic peptides, and LTB_4 in neutrophils.[188] When rabbit neutrophils were treated with phorbol-12-myristate-13-acetate (PMA), a protein kinase C activator that stimulates aggregation, degranulation (of specific granules), and the oxidative burst, they were found to be nonresponsive (within 3 min) to FMLP and LTB_4 stimulation. Parameters measured were inhibition of degranulation, *N*-acetyl-beta-glucosaminidase release, sodium influx, lysozyme release and increase in intracellular free calcium concentration upon stimulation with the FMLP or LTB_4. This inhibition was observed without interference of binding of FMLP or LTB_4 to their respective receptors. This inhibition may have resulted from altering the microenvironment of the receptor or phosphorylating the receptor by activation of protein kinase C (PKC), which could in turn inhibit the signal process (production of inositol-1,4,5-triphosphate (IP_3) or others) that leads to the mobilization of calcium. LPS primes macrophages for these lipid-mediated events, but does not readily initiate them. The arachidonic acid release associated with this process involves ligand-receptor binding, receptor clustering, sodium ion influx, synthesis of a rapidly turning-over protein, and calcium ion influx.[196]

These observations lead to the next step: looking for activation pathways that modulate other activation pathways. Platelet-derived growth factor (PDGF) is an excellent example of a multiple pathway activating agent, with positive and negative effects on various activation pathways.[397] When it binds to its receptor, it initiates a tyrosine kinase cascade, sodium/potassium ATPase activity, and sodium/hydrogen ion antiport activity. It also causes release of calcium ions from intracellular stores. This process may be a result of IP_3 and diacylglycerol (DAG) production by phospholipase C (PLC). The DAG and calcium activate the PKC, which phosphorylates the epidermal-derived growth factor (EDGF) receptor, decreasing its ability to bind EDGF. The arachidonate released by the PLC is converted to PGE_2 by cyclooxygenase. This autocrine binds to its own receptor, as described earlier, and activates an adenyl cyclase. Also, the arachidonate inhibits the guanosine triphosphatase activating protein (GAP) from binding to the G-protein associated with the activation of the PGE_2 receptor and its adenyl cyclase. At the same time, the arachidonate, phosphatidic acid, and phosphatidylinositol phosphates released by phospholipase activity prevent GAP binding to c-Ha-ras protein, allowing it to remain activated for growth stimulation. However, this inhibition of GAP binding feeds back onto the PDGF receptor, which shows enhanced binding of GAP as part of its signal transduction.[398] Therefore, the free arachidonate is part of a negative feedback on the PDGF activated receptor. Also, this binding of GAP may free ras to stimulate growth.

When the GAP binds to the PDGF receptor, the tyrosine kinase activity of the receptor is activated and the bound GAP and two tyrosine residues in the receptor (tyrosines 751 and 857) undergo phosphorylation.[398] These latter autophosphorylations are necessary for GAP binding. Binding of PI_3 kinase activity and three other proteins to the PDGF receptor requires PDGF and kinase activity. Arachidonic acid release mediated by receptor activation may have positive effects on metabolism (catabolism), but other negative effects on growth stimulation. For instance, alpha-interferon induces the transient activation of phospholipase A_2 in 3T3 fibroblasts.[399-401] The phospholipase inhibitor bromophenacyl bromide specifically inhibits the interferon- induced binding of nuclear factors to the interferon-regulated enhancer element (ISRE). This arachidonate activity was associated with an oxidized metabolite of arachidonate but not one produced by cyclooxygenase or lipoxygenase, because inhibitors of these enzymes enhanced the ISRE binding and associated gene expression. Oddly enough, treatment of c-H-ras-transformed 3T3 cells with alpha- and beta-interferon led to a stable reversion of transformation phenotype. Therefore, the unspecified oxidation products of arachidonic acid triggered by alpha- and beta-interferon may bind to the GAP of the H-ras complex, altering its function or facilitating its downregulation of c-H-ras growth stimulation (since native arachidonate would be expected to stimulate the G-protein growth effects).[199,401-403]

A similar stimulation of GAP binding to tyrosine kinase associated with a receptor has been seen with the high-affinity receptor for human nerve growth factor (NGF).[404] In PC-12 cells transfected with the human Trk protooncogene, that codes for a 140-kD M_r membrane tyrosine kinase associated with the high-affinity NGF receptor, both GTP/GDP exchange factor and GAP are activated by NGF.[405] However, GAP tyrosine phosphorylation is not activated as in PDGF receptor and EDGF receptor interactions. The Trk kinase increases tyrosine phosphorylation of cellular proteins in general and activates several serine/threonine kinases in particular. Either GAP is indirectly phosphorylated by these cytosolic kinases and its activity is altered, or the ras/GTP-binding/hydrolysis/GDP-release cycle is part of the transduction pathway for NGF.[404,405] Such a cycle could be important in the cytosolic factor activation of NADPH oxidase membrane components (cytochrome b_{558}).

The ubiquitous, and therefore overlapping, central role of ras gene products in cellular transduction and signaling pathways has been recently clarified somewhat.[402] In all cases to date of ras-mediated responses, there appear to be four components involved: (1) the ligand/antiligand complex (occupied receptor), or the sensor, in the case of physical or nonspecific chemical stimuli; (2) an adaptor connecting the other components to the plasma-membrane receptor; (3) a guanine nucleotide-releasing factor bound to the adaptor or to the ras itself; and (4) the GAP associated with the ras. Besides the receptors already discussed, the CD_3/T-cell-receptor activation pathway also has this basic design.[406] However, this pathway is not the only one activated in T cells during immunostimulation.

T cell mitogen or promoter stimulation (i.e., phytohemagglutinin, Con A, or PMA treatments) leads to phospholipase A_2 and lipoxygenase activation. These initial events lead to guanylate cyclase stimulation, subsequent rises in levels of cGMP, and DNA synthesis. However, cyclooxygenase and thromboxane synthase activities yield no such outcome. Both phospholipase A_2 and arachidonate, when added extracellularly, stimulate guanylate cyclase. Treatment with 5,8,11,14-eicosatetraynoic acid inhibited the guanylate cyclase stimulation.[407] This eicosinoid inhibits both cyclooxygenase and lipoxygenase. Indomethacin, which is specific for cyclooxygenase, had no such effect. Phospholipase A_2 from hog kidney and snake venoms such as that of *Vipera russelli* caused activation of membrane and soluble magnesium ion-dependent guanylate cyclase (4.5- and 3-fold for membrane and soluble enzymes, respectively, with 2.5 µg/ml of the kidney enzyme). Phospholipase inhibitors such as dibutyl cAMP at 0.5 mM greatly inhibited guanylate cyclase activation. When added to solublized forms of human peripheral blood lymphocyte guanylate cyclase, they could still be activated with arachidonate. When added to intact cells, the arachidonate stimulated only the membrane guanylate cyclase (peak effect at 1 µM).

5-, 11-, 12- or 15-Hydroperoxyeicosatetranoic acid (HPETE) activated spleen cell guanylate cyclase whereas monohydroxy derivatives that were not oxidatively active did not. 5-HETE, like arachidonate, did activate intact lymphocytes two- to sevenfold in guanylate cyclase activity and were concluded to do so by activating receptors for lipid peroxidation activation. Lipoxygenase inhibitors inhibited this effect. These nonperoxy compounds activate phospholipase and arachidonate metabolism that then produces the lipid peroxides. 15-HETE can stimulate guanylate cyclase. It is a better inhibitor of 5- and 12-lipoxygenase but not the 11- and 15-lipoxygenase or cyclooxygenase. 15-HETE does, in fact, inhibit mitogenesis in mouse spleen cells and is immunosuppressive.

Some of the confusion about mitogen stimulation pathways and phorbol effects can be cleared up when the lipoxygenase/cGMP pathway is also included in the T lymphocyte-activation scheme. PHA stimulates predominantly the membrane guanylate cyclase while PMA stimulates primarily the cytoplasmic soluble form. Neither mitogen demonstrates direct effects on the respective enzyme forms. Both mitogens show calcium-dependent guanylate cyclase and DNA synthesis stimulation prevented by phospholipase A_2 inhibitors such as quinocrine and by lipoxygenase inhibitors such as nordihydroguaiacetic acid (NDGA).[277] PMA stimulation of guanylate cyclase is not inhibited by inhibitors of phospholipase A_2 and lipoxygenase, whereas PHA stimulation is. Hydroxyl radical production has been reported to be involved in guanylate cyclase (soluble enzyme) stimulation by PMA, which is inhibited by 15-HETE.[408]

These observations not only open up the possibility of dual-pathway stimulation by multiple agents, but also the importance of timing and sequence of such stimulation. When CD+4 and CD+8 T cells from HIV-infected individuals were stimulated with antibody against the CD3 surface marker, they underwent programmed cell death, apoptosis, instead of the usual mitosis.[409,410] Apoptosis is marked by degradation of chromatin into discrete fragments that are multiples of approximately 190 DNA base pairs, dilation of the endoplasmic reticulum, plasma membrane blebbing, and foamy cytoplasm.[411] This effect has been associated with increased production of TNF and transforming growth factor beta (TGF-beta) in AIDS patients. TNF has been shown to stop cell progression in the cell cycle, a possible mechanism for LPS-induced ionizing radioresistance.[366]

TGF-beta$_1$ and TGF-beta$_2$ cause 50% inhibition of activated- macrophage production of hydrogen peroxide at 0.6- and 4.8-pM concentrations, respectively, after two days of incubation with the TGF-beta before activation.[244,412] This suppression could be overcome by treatment with gamma-interferon, TNF-alpha, or TNF-beta. TGF-beta was not toxic to the macrophages when measured by trypan blue exclusion. Although the TGF-beta did not affect phagocytoses, it did decrease mouse macrophage adherence to plastic. Major sources of TGF-beta are degranulating platelets, endothelial cells, fibroblasts, keratinocytes, tumor cells, and antigen-stimulated T cells. Macrophages will produce TGF-beta as an autocrine when treated with LPS. P815 mastocytoma cells produce another factor, macrophage deactivating factor (MDF), that increases the K_m of NADPH oxidase for NADPH. MDF is immunologically and functionally distinct from TGF-beta. TGF-beta also suppresses T and B cells. Although the mechanism of inhibition of NADPH oxidase by TGF-beta is not understood, TGF-beta is known to increase mRNA for IL-1-alpha and -beta and TNF-alpha by enhanced transcription of these genes.[413] However, the translation of these mRNA's into protein is blocked. TGF-beta inhibits the induction of IL-1-beta and TNF-alpha by LPS. However, PMA and PHA induction of TNF-alpha and IL-1-beta is not affected by TGF-beta. It also inhibits gamma-interferon production by PHA-activated mononuclear cells. Neither cAMP nor PGE_2 is induced by TGF-beta. Therefore, this induction cannot be its mechanism of interaction in inactivation of macrophages. Mouse macrophages from wound sites produced TGF-alpha mRNA and TGF-alpha protein. Macrophages in culture treated with modified (acetylated) low-density lipoprotein or LPS produced TGF-alpha transcripts within 3 hours, and they remained high for 6–9 hours but disappeared by 48 hours. TGF-alpha mediates epidermal regrowth, angiogenesis, and formation of granulation tissue. Therefore, macrophages have a cytokine feedback control that stimulates or suppresses cell growth and NADPH oxidase activity.

Heat shock proteins produced by leukoctyes and other cells when quiescent or under stress are associated with inhibition or enhancement of their growth. In progesterone-sensitive cells, this growth inhibition may be released by the progesterone, and in lymphocytes this release is accomplished by mitogen or antigen stimulation.[414,415] The artificial steroid RU486 has a high affinity for the cytoplasmic progesterone receptor and inhibits the actions of progesterone.[414] The progesterone receptor is attached to HSP90 and the nuclear protein p59. When progesterone enters the cell, the HSP90 and p59 are released, allowing the receptor to bind the hormone-responsive element of the genomic DNA. Transcription factor then binds and transcription is initiated. When RU486 is present, the HSP90 and p59 are not released or, if released, the transcription factor does not bind to the DNA. RU486 not only inhibits the synthetic machinery of cells dependent on progesterone, but also increases prostaglandin (PGF_2-alpha) production, causing increased myometrial contraction and dilation and softening of the uterine cervix. Endometrial loss then follows RU486 treatment, resulting in spontaneous abortion.

HSP70 mRNA is elevated in circulating human peripheral blood lymphocytes, which are in the G_0 state.[415] When incubated with serum, PHA or in culture medium, the HSP70 mRNA decreases within 30 min after such conditions start. With this decrease, there is an increase in c-fos mRNA. This constitutive high level of HSP70 mRNA has been proposed to be a carryover from G_2 accumulation.[415] Stimulation could result from restarting the cycle only, although the c-fos is usually associated with proliferation. In A431 cells stimulated with epidermal growth factor and P12 cells stimulated with nerve growth factor, it is associated with cell differentiation and not DNA synthesis. The lymphocytes may have entered the G_1 state without the S phase. In HeLa cells and human embryonic kidney cells (293), after serum starvation and reintroduction of serum, there is a transient increase in HSP70 mRNA between 12 and 18 hours after serum stimulation to 10-fold, the starvation level.[416] This elevation was accounted for by a 10- to 15-fold increase in HSP70 gene transcription.[416] Evidence has been presented for the linkage of this stimulation to DNA sythesis in HeLa and 293 cells. The HSP70 gene has been shown to have serum-controlled transcription regulatory sequences.[415] Besides serum and heat shock responsiveness, the HSP70 gene is inducible with adenovirus 5 EIA.[417] However, the serum induction is independent of EIA, suggesting more than one regulatory element on the HSP70 gene.[417] The transcription of the sequences of the adenovirus pre-early region increases in 293 cells with serum starvation and decreases when HSP70 transcription is maximal. HSP70 appears to be essential to cell growth, and in the lymphocytes, it places the cells in the ready state for DNA synthesis.[415] This growth function may be a manifestation of heat shock protein chaperonin function for newly synthesized proteins. This function as stated above would be involved in both differentiation without DNA synthesis and as a prerequisite to DNA synthesis because of the need for protein synthesis and delivery to sites of proteins needed for DNA synthesis.

Even more primitive compounds that interact with proteins following oxidative modification (the primitive analogue of stress proteins interacting with oxidatively modified proteins) may be involved with immune responses, cell-to-cell interactions, and nonself recognition.[418] In insects (i.e., the wax moth *Galleria mellanolla*), the activation of prophenoloxidase by microbial products (endotoxin) yields melanin coating on the microbes. Hemocytes (called granular cells), when stimulated by bacteria or endotoxin, in the hemolymph of moths, rapidly degranulate and induce hemolymph melanization and coagulation. The plasmatocytes, which form lamellipodia and have very few granules, showed a sixfold increase in phagocytosis following stimulation of cocultured granular cells with laminarin or endotoxin. However, anticoagulants inhibit the endotoxin-stimulated prophenoloxidase system, but not the laminarin-stimulated one. The prophenoloxidase activation resulted in melanin coating of target cells that enhanced recognition and phagocytosis by plasmatocytes. Serine proteases activate the prophenoloxidase by proteolysis of the proenzyme.

Melanin deposition around parasites is a common mechanism associated with humoral and cellular immune defenses of arthropods. Even though insects and other invertebrates do not produce immunoglobulins as recognition molecules, they show high selectivity and reactivity toward nonself cells, that is, parasites and microbes. Oxidative products may be essential in these cases, similar to stress proteins recognizing oxidized or denatured proteins and heme oxygenase recognizing heme freed from denatured hemoproteins. In humans, peroxidases in the brain (substantia nigra), in the inner ear, and in other peroxidase- containing cells (such as granulocytes and monocytes) appear to be capable of significant melanin production.[154,336,338,419-422] These are tissues that do not become exposed to UV radiation, and therefore, where melanin would serve no purpose as a UV protector. The autoxidation of dihydroxydopamine, the precursor of melanin, forms the hydrogen peroxide that drives melanin production by peroxidase without additional oxidase (especially under alkaline conditions) in these tissues and cells.[423] One can only speculate on the possible immunologic functions of these tissue melanins in humans.

IV. OTHER ENZYMES SIMILAR TO NADPH OXIDASE

A. GALACTOSE OXIDASE

D-Galactose oxidase from the fungus *Dactylium dendroides* is a nonheme metalloprotein of 55-kD M_r, with one copper atom per molecule as the only cofactor.[424] In the resting enzyme, most of the copper ions are in the Cu(II) state (based on EPR data). The EPR signal intensity is maintained in steady state throughout the oxidation of D-galactose at the methyl hydroxy position (6) to the corresponding aldehyde and hydrogen peroxide. Superoxide dismutase inhibits the rate of D-galactose oxidation and increases slightly the Cu(II) EPR signal. Ferricyanide or superoxide activates the enzyme fourfold or greater and decreases the EPR signal to zero. These data indicate that the enzyme oscillates between Cu(I) and Cu(II), with superoxide bound to the Cu(II) state as a short-lived intermediate. The Cu(III) is the initial oxidant that converts the galactose alcohol to the aldehdye and is oxidized by molecular oxygen. Trace transition metals cause variation in the enzyme activity by acting as nonenzymatic scavengers of superoxide by disproportionation. The removal of free superoxide from the galactose oxidase leads to an inactive enzyme. This effect bears some similarity to superoxide production by green hemoprotein, where hydroxyl radical production (evidenced by self-bleaching) can be inhibited by removal of superoxide. The resting enzyme has a mixture of catalytically active, Cu(III), and inactive, Cu(II), forms, again reminiscent of GHP, where iron is probably in a mixture of Fe(II) and Fe(III). Generally, 300 turnovers of galactose oxidase occur before one superoxide molecule leaks out to make an inactive form of the enzyme. If this superoxide is rapidly removed (as with SOD), then the enzyme will eventually completely inactivate, following a lag created by the turnovers. The ferricyanide oxidizes the Cu(II) to Cu(III), reactivating the enzyme. Therefore, the Cu(II) intermediate that is active has bound superoxide, and the enzyme normally cycles between the Cu(III) and Cu(I) forms with Cu(III) oxidizing the sugar and Cu(I) reducing the oxygen and forming the superoxide. In the resting enzyme, the inactive form dominates, being at 75% of the total enzyme. Essentially, the Cu(II)/superoxide complex normally yields Cu(III) and hydrogen peroxide in the acidic microenvironment generated by the oxidation of the galactose alcohol to aldehyde. This reaction is essentially a superoxide dismutase reaction. The main difference between this superoxide producer and NADPH oxidase is the difficulty in which it regenerates the catalytically active form after superoxide loss. This difficulty results from the fact that unlike Fe(II), Cu(II) does not readily oxidize in the presence of oxygen.

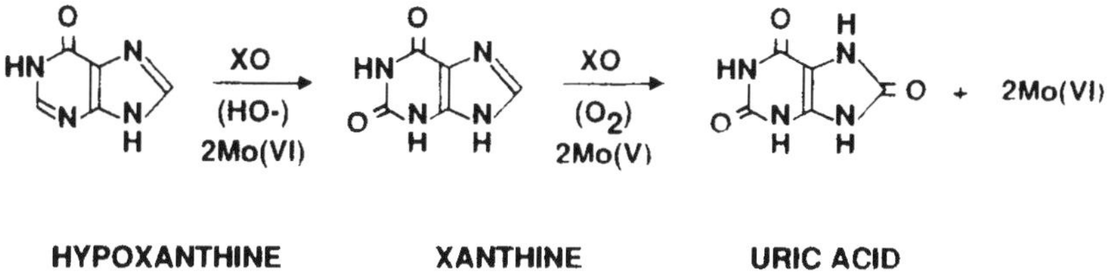

FIGURE 10. Reactions catalyzed by xanthine oxidase (XO). The co-factor molybdenum (Mo) undergoes cyclic reduction and oxidation.

B. XANTHINE OXIDASE

Xanthine oxidase is a 275-kD metalloflavoprotein containing two flavin adenine dinucleotides (FAD), two molybdenum ions, and eight nonheme iron ions.[425] It can oxidize various aldehydes to carboxylic acids with the removal of two hydrides (two protons and four electrons). Xanthine oxidase can oxidize hypoxanthine to xanthine by 2-hydride removal and can further oxidize xanthine to uric acid by oxidation at the 8 and 9 positions (see Figure 10). Strangely enough, in provinces of China where nitrosamines are high in fermented foods and molybdenum is deficient in the soil and agricultural products, esophageal and stomach cancer is inordinately high in poultry and people that eat these foods.[426] This effect contradicts the idea that free radicals cause genetic damage, which in turn leads to cancer.

The mechanism of xanthine oxidase action is reminiscent of mitochondrial and chloroplast redox chains. Electrons are passed from the molybdenum atoms to the flavin to the iron and oxygen. The iron is bound not to heme but to sulfur like the nonheme iron proteins of mitochondria or chloroplasts. Furthermore, the enzyme contains a persulfide (S–S–) group essential for activity. Figure 10 shows one proposed mechanism for electron and proton separation seen in the redox chain of xanthine oxidase where the initial event is the 2-electron reduction steps that involve two 1-electron reductions of molybdenum (IV) to molybdenum (V). This mechanism has a strong resemblance to the oxygen-generating photosystem of photosynthesis and to the nitrite-producing system of nitrate reductase (also a molybdenum-bearing enzyme but with a cytochrome b iron-containing protein).

Xanthine oxidase has been proposed as a major source of reperfusion injury in tissues reoxygenated following ischemia.[427] The tissues in which xanthine oxidase is implicated as the cause of reperfusion injury include small intestine, stomach, liver, pancreas, skeletal muscle, kidney, heart, brain, and skin.[427] However, evidence exists that the proteolytic conversion of xanthine dehydrogenase to xanthine oxidase is an early event of autolysis.[428,429] Furthermore, the influx of neutrophils and their adherence to blood vessel walls in reperfused areas (especially in cardiac tissue) confounds determination of the contribution of xanthine oxidase to peroxidative reperfusion injury.[430,431] Green hemoprotein and damaged redox chains (mitochondria) could also contribute to the reperfusion injury. Furthermore, the proteases released by inflammatory cells and damaged tissue may yield measurable but nonphysiologic xanthine oxidase activity (after-the-fact activation).[429] The best evidence for the participation of xanthine oxidase in reperfusion injury of an ischemic tissue is found in the intestines.[428]

The main effect of ischemia in the small intestine is increased capillary permeability that results in edema and loss of fluid into the lumen of the small intestine.[428] Substances, other than reactive oxygen intermediates, that have been implicated in increased vascular permeability associated with ischemia are histamine, prostaglandins, lysosomal enzymes, and bacterial endotoxins. However, their role has not been borne out by inhibition studies with antihistamines, indomethacin, or methylprednisolone.[432] Also, the magnitude of effect in relation to dose necessary for these agents to create the effect show a nonphysiologic relationship.[432] Evidence for reactive oxygen intermediate participation in reperfusion injury is in the significant inhibition of vascular permeability with pretreatment with superoxide dismutase, catalase, and/or dimethylsulfoxide (a hydroxyl free radical scavenger).[288,432-437] The participation of xanthine oxidase in the production of these oxygen products is supported by the fact that allopurinol, a specific competitive inhibitor of xanthine oxidase, protects from reperfusion damage as well as reactive oxygen scavengers. However, allopurinol does produce other metabolic and physiologic effects unrelated to its inhibition of xanthine oxidase. Other specific inhibitors of xanthine oxidase, which do not have these other effects, also protect against ischemia-reperfusion injury. Pterin aldehyde, an oxidative product of folic acid found in commercial preparations of the vitamin, is 1000 times more potent than allopurinol and about equal in potency to oxypurinol, a metabolic product of allopurinol in xanthine oxidase inhibition. Another piece of strong evidence for the role of xanthine oxidase in oxidative injury in the small intestine is the fact that the small intestine has the highest concentration of xanthine oxidase in the body, especially in the mucosal layer.[432] The tip of each mucosal intestinal villus has higher activity than its base, which corresponds to the tissue sensitivity to ischemic injury. There is also a gradient of xanthine oxidase content from the duodenum at 10 U/g protein to the ileum at 3 U/g protein. The complete conversion from the xanthine dehydrogenase (NAD-dependent enzyme) to xanthine oxidase requires only 1 min of ischemia. The heart requires 15 min, and other tissues require 1 hour or more for the conversion. Evidence has been presented previously, based on serine protease inhibitor and calcium-calmodulin inhibitor studies, that shows the conversion is mediated by a calmodulin-regulated intracellular serine protease.[428] Ischemic damage to the calcium ATPase pump, by energy depletion or previous reactive oxygen damage, could accelerate the rate of conversion of the dehydrogenase to the oxidase.[329] The substrate level for xanthine oxidase is 20 to 40 µM of hypoxanthine in normal bowel, which rises to more than 400 µM during ischemia. However, the K_m of xanthine oxidase for hypoxanthine is only

11 μM; therefore, the enzyme is easily saturated and oxygen is actually the limiting substrate in the tissue (K_m of oxygen is less than 20 mmHg). Oxygen tension in tissue transiently exceeds the normal tissue level of 20 mmHg and may reach 70 mmHg. Ischemia-reperfusion in the intestine as well as in cardiac tissue induces neutrophil influx. This influx can be prevented by treatment with SOD or allopurinol, and it has been shown that treatment of albumin with superoxide results in the production of a neutrophilic chemotactic agent.[433,434] The oxidation of unsaturated fatty acids carried by plasma albumin may be responsible for the chemotactic activity. Once the neutrophils have entered the tissue, the NADPH oxidase and lipoxygenase activities of the leukocytes can contribute to the vascular permeability changes.[433,435,436]

Whereas various combinations of SOD, catalase, hydroxyl radical scavenger, and allopurinol produced protective effects in the tissues described other than intestine, the results were not always as good or clean-cut. Skeletal muscle responses were the closest to small intestine with kidney being next. Allopurinol gave more equivocal results in cardiac muscle ischemia. Also, allopurinol treatment of mouse granulocytes inhibits their superoxide production. This fact alone casts doubt on the specificity of allopurinol for xanthine oxidase. Brain tissue studies present the least convincing evidence for xanthine oxidase participation in reperfusion oxidative injury of the brain because brain has the lowest content of xanthine oxidase in the body. However, abundant evidence from studying tissue levels of natural antioxidants (i.e., ascorbate, alpha-tocopherol, reduced glutathione, and ubiquinone) strongly suggest a role for free radical damage in central nervous system (CNS) ischemic injury. What is clear is that reactive oxygen products are central in ischemia-reperfusion damage to tissues and organs of the body, regardless of the source of free radicals, and that there are several probable sources, dependent on the tissue, that may complement each other in effect.

C. CROSSLINKED GLUCOSE OXIDASE AND PEROXIDASE

The antimicrobial activity and tumor cell cytoxicity of oxidase- and peroxidase-based systems *in vitro* are well documented. Therefore, it was reasonable to assume that an oxidase/peroxidase system could be constructed as an antitumor agent for *in vivo* use. Because of convenience and availability, the fungal flavoprotein enzyme glucose oxidase (from *Aspergillus niger*) and the plant peroxidase horseradish peroxidase were chosen.[315] However, glucose oxidase does not normally produce superoxide, making it unlike NADPH oxidase, and horseradish peroxidase cannot oxidize chloride to hypochlorous acid, making it unlike myeloperoxidase.[231,315] The toxicity of hypochlorous acid produced by myeloperoxidase has been attributed to the formation of chloramines from amino acids and proteins.[69,82,145,437-441] Therefore, at best, this pair could be expected to be a poor mimic of the oxidase/peroxidase couple of leukocytes. However, glucose oxidase like NADPH oxidase has a FAD binding site, and horseradish peroxidase like cytochrome b_{558} has protoporphyrin IX noncovalently bound into its active site. To keep the enzymes in proximity in order to maximize the peroxidative effects, they were cross-linked with glutaraldehyde and serum albumin.[315] Even though weaker, nonspecific cytotoxicity was expected with glucose oxidase and horseradish peroxidase, they were used. Very malignant Novikoff hepatocarcinoma (including metastases) in rats, spontaneous mammary adenocarcinomas in rats, L1210 leukemia in mice, B16 melanomas in mice, and carcinogen-induced squamous cell carcinomas in hamster cheek pouches all responded to regional or local injections of the cross-linked enzymes in saline and glucose solution.[315,340,442-444]

The Novikoff hepatocarcinoma responses were 50 to 100% cures in animals treated.[315,444] Lactoperoxidase could be substituted for horseradish peroxidase, and similar results were obtained.[110,111,315] However, galactose oxidase substitution or the absence of either the glucose oxidase or the respective peroxidase in the preparation failed to bring about tumor regression.[315] L1210 leukemia in outbred white mice showed 50% cures. Treatment of B16 melanoma prolonged life from 23 days postinoculation of controls to 37 days with treatment. However, no simple dose-response curve was demonstrated: 1.25 mg of cross-linked preparation per kilogram of body weight worked as well as 160 mg/kg.[444] Also, smaller squamous cell carcinoma tumors responded better (some with cures) than did larger cheek pouch tumors, which did not respond at all.[442,443] Hyperbaric oxygen provided the same response in cheek pouch tumors as the cross-linked enzyme, and the two treatments were neither synergistic nor additive.[442,443] No direct injury to normal tissue was observed in histopathological sections of treated animals.[110,111,315] However, some animals with heavy tumor loads (P388 lymphoma, L1210 leukemia, some with Novikoff hepatocarcinoma) died more quickly when treated than did untreated controls due to an apparent "necrotic crisis" from dying tumor tissue.[110,116,315]

Although ABTS-stainable particles were recovered from omentum of treated animals up to a week after treatment, the lack of a dose response and limited tumor contact (particularly in Novikoff hepatocarcinomas that metastasized to regional lymph nodes) suggested an alternative mechanism to direct cytoxicity. Like the studies of antigen effects on splenic peroxidase and anamnestic splenic chemiluminescence in immunized mice, the rats showed increased splenic peroxidase following inoculation with Novikoff tumor cells.[340] In untreated rats, the peroxidase level began to rise at 4 days, peaked about 8 days after inoculation, and declined to below baseline as the tumors progressed (most animals died between 10 and 12 days postinoculation). In comparison with mice, when nontumor antigens were used, the splenic peroxidase levels began to rise at 2 days postinoculation and peaked at 4 days postinoculation.[340] Animals treated with cross-linked oxidase/peroxidase showed a rise in splenic peroxidase beginning at about 4 days and peaking at 12 to 17 days postinoculation, well above that of untreated animals and remaining above baseline beyond 20 days. Also, the baseline tritiated thymidine uptake (a measurement of DNA synthesis) was much higher in tumor-bearing animals, but the stimulation indices with Con A stimulation vs. control T cells dropped from 15 to 2, with the major decline between 2 and 5 days postinoculation.[231,315,340] In treated animals (treated for 4 days, starting 3 days after tumor inoculation), the stimulation indices dropped to about 2 on day 12 and then began to rise thereafter. Con A (5 μg/ml) treated rat T cells showed a 80–99% inhibition of DNA synthesis when treated with mitomycin-C-inhibited Novikoff hepatocarcinoma cells.[340] After *in vitro* treatment of these hepatocarcinoma cells with the cross-linked oxidase/peroxidase for 6 hours, the Novikoff cells inhibited the T cell DNA synthesis 0 to 50%.[340] Therefore, the oxidase/peroxidase was modifying the Novikoff cells so that the immune cell response to them was also altered. By comparing the effective and noneffective splenic T cell responses of the rats to tumor development and their peroxidase levels, one can find that peroxidase rises precede changes in T cell responsiveness.[110,111,315,340] Therefore, the peroxidative modification of complex cellular antigens modified the immunologic response to the tumor. Other data exist that support the concept that the immunostimulation by oxidase/peroxidase systems precedes development of specific cellular immunity.[274,280]

When the feeder-layer phenomenon of macrophage (RAW 264.7 mouse cells) support for growth of CTLL-2 mouse cytotoxic T cells was examined, the growth stimulation was found with certain dilutions of nanometer-sized particles of cross-linked glucose oxidase-horseradish peroxidase.[397] Using 2-mercaptoethanol/IL-2 maximally supported CTLL-2 growth as the standard, RAW 264.7 cells treated with 1/2000 dilution of 1 mg/ml particles showed a peak growth support of 47.1% of the mercaptoethanol supported growth. When P388D$_1$ mouse macrophage cells were substituted for RAW 264.7 cells and treated with oxidase/peroxidase particles at 1/1000 dilution, only 12.4% of the maximum growth was supported compared to 24.2% for RAW's. The P388D$_1$ cell line showed no support for CTLL-2 cells when treated with oxidase/peroxidase particles at 1/2000 and 1/4000 dilutions and –12.8% (inhibition) growth support without oxidase/peroxidase particles. Therefore, growth support of cytotoxic T cells stimulated by cross-linked oxidase/peroxidase was linked to a response of the macrophages. The cross-linked oxidase/peroxidase was shown to moderately enhance thiol production by macrophages as well.[397] Consequently, the antitumor activity was far more complicated than direct cytotoxicity of reactive oxygen intermediates.

The link between oxidative/peroxidative activity of the immune response and facilitation by artificial oxidase/peroxidase systems has been extended beyond nonpathologic antigens and tumor cells to viruses.[445] When Swiss (ICR) mice were infected for less than 24 hours with the picornavirus D variant of encephalomyocarditis virus (EMCV-D), alterations in the immune systems and cytokine levels of the mice were observed. Whereas five 1-mg doses of glucose oxidase/peroxidase produced levels of interferon-alpha and -beta in spleens of infected mice comparable to controls, there was a marked increase of interferon-alpha and -beta in spleen cells from mice infected with the virus and treated with lower doses of the insoluble cross-linked cytotoxic oxidase/peroxidase (ICCOPS; five 0.1-mg doses). The enhanced production of interferons-alpha and -beta occurred for 16 hours, and the gamma-interferon first appeared at the end of the 16-hour postinoculation period. In uninfected mouse spleen cells, ICCOPS at doses ranging from 0.0025 to 10 mg/3.74×10^6 cells produced no interferon. If cells from female mice were exposed to 0.02 mg ICCOPS/3.75×10^6 spleen cells, where 30% of the cells showed cytotoxicity, high levels of alpha- and beta-interferons were produced at 16 and 20 hours postinoculation. In these mice, by 24 hours postinoculation, 60% of the interferon was gamma-interferon. At a dose of 0.2 mg/3.75×10^6 spleen cells, where cytotoxicity was 50%, no interferon was produced. Spleen cells from highly responsive female mice produced TNF at 4 and 8 hours, when treated for 5 days with 0.1 mg ICCOPS per mouse per day, after infection, but low responders or male mice did not produce TNF.

Also, the optimal level of ICCOPS (0.1 mg per mouse per day) yielded marked increases in the Thy 1.2+ spleen cell population instead of the usual large decrease at 24 hours after infection with EMCV-D. Therefore, the ICCOPS either stimulated protection of the T cells or changed their trafficking patterns. In uninfected female mice, this ICCOPS concentration increased expression of Class I major histocompatibility (MHC; H-2) antigens, while when infected with virus, the ICCOPS enhanced the H-2 antigen production by spleen cells of both sexes (probably an indirect effect due to enhanced interferon production). Again, the peroxidative activity affects the immune response toward virus, but only in infected cells. Therefore, oxidase/peroxidase products influence immune cells in certain states to yield various cytokines that then engage the specific immune response.

In addition, the endogeous oxidative activity of macrophages can be enhanced by peroxidase.[446,447] Soluble horseradish peroxidase, when given to mouse peritoneal macrophages, enhances their phagocytic and oxidative activity, even though the peroxidase has been shown to be proteolyzed at least in part.[117] Porphyrin does not provide this enhancement.[448] These observations may indicate that the noncovalently bound heme of peroxidase is transferred to apocytochrome b of the NADPH oxidase. Whether heme from cross-linked oxidase/peroxidase could be salvaged in this manner is unknown.

V. SUMMARY

NADPH oxidase is a flavocytochrome attached to regulatory and stabilizing subunits. Its activation proceeds by several possible phosphorylating pathways, involving G-proteins, calcium, IP_3, DAG, cAMP, arachidonic acid, cyclooxygenase, lipoxygenase, and tyrosine phosphorylation. Dominant pathways seem to be determined by the cell type or state of the cell in the case of macrophages. Essentially, NADPH oxidase can be viewed as a type-b cytochrome fed electrons from a flavoprotein and NADPH oxidase components, which are modified by cytoplasmic phosphorylated and phosphorylating components. Of the type-b cytochromes discussed in this book, it is the most powerful producer of superoxide without self-destructing from autoxidation. However, its stability and reactivity with reducing sources is wholly controlled by the other ancillary components of the oxidase.

In cell types other than granulocytes or macrophages, such as B lymphocytes, hepatocytes, HeLa cells, etc., the cytochrome b component (22 kD) is constitutively produced, but is incapable of stable, sustained superoxide production. This fact raises the question as to its function. The linkage of heme oxygenase (a stress protein) induction to membrane UV or oxidative damage suggests a role for a membrane-bound or integrated hemoprotein. The 22-kD type-b cytochrome component of NADPH oxidase could be such a hemoprotein. Not only is heme oxygenase induction associated with membrane UV damage, but also the activation of H-ras and, subsequently, NF-kappa-B is associated with such damage. The latter is an important transcription factor for the expression of many cellular genes and latent retrovirus, including IL-2 and HIV, respectively. The fact that NADPH oxidase activity is associated with upregulation of cytokines such as TNF and gamma-interferon and downregulation of cytokines such as macrophage deactivating factor and transforming growth factor suggests a much more important role for nonphosphorylating redox activity than just cell or microbial killing. Whereas bacterial and eukaryotic cells may have originally responded to oxidative damage by developing induction of antioxidant enzymes and antioxidant synthesis, the immune system response and cell growth regulatory responses imply much more. The antioxidant defense strategies of cells have evolved into anticipatory specific immune and growth and repair responses. In fact, responses to cytokines (such as TNF and IL-1) have evolved to mimic the original nonspecific oxidative stimuli. In a sense, they represent a primordial "phantom" oxidative insult that no longer exists but was and remains an evolutionary driver for cell survival (defense, repair, growth, and immune responses).

The unstable type-b cytochrome is stabilized by heat shock proteins and heat shock-like components of NADPH oxidase so that it does not have to evolve its own internal stabilization and yet remain plastic and responsive to regulation and environmental effects when necessary. In a sense, the NADPH oxidase evolved from a primordial redox (ionizing and UV radiation) membrane sensor that amplifies the production of reactive oxygen species without self-destruction during the autoxidative process (unlike the green hemoproteins). Even the destructive inflammatory process of NADPH oxidase is mimicked in tissue remodeling and the cytotoxic lymphocyte responses mediated through programmed cell death. Here, the cells utilize signals that lead to breakdown of proteins and nucleic acids prior to disrupting the cell by phagocytosis in order to prevent initiation of a true inflammatory response. This review of the oxidative/peroxidative effects on the immune system suggests the hypothesis that the specific immune response evolved in order to suppress or counter redox injury (originally from radiation). It further suggests that organisms continue to use redox injury (even self-imposed) not only to destroy invading organisms, altered self and unwanted redundant cells, but also to drive cellular responses to nonoxidative or even noninjurious stimuli.

Chapter 4

NITRIC OXIDE SYNTHASE

CONTENTS

I. WHY NITRIC OXIDE SYNTHASE HAD TO BE A CYTOCHROME B

In Chapter 3, the cytotoxic enzyme NADPH oxidase and enzymes like it, which are capable of forming reactive oxygen intermediates, were shown to be capable of initiating some synthetic processes. However, except for the induction of antioxidant enzymes and antioxidant synthesis, the responses are rather nonspecific and can be mimicked by other denaturing processes, nonenzymic oxidations, and hormone and cytokine actions. However, there is a family of redox enzymes, nitric oxide synthases (NOS), that have evolved into primary intercellular communicators, with nitric oxide being the primary chemical mediator of this communication. Although nitric oxide synthases are all capable of cytotoxic action and are frequently used for this purpose by leukocytes, their products, nitric oxide and its derivatives, have specific transduction pathway activities.[60,449,450]

But why did nitric oxide synthase have to be a cytochrome b? First, NO has a high propensity for binding to ferrous heme.[70] Its NO-binding equilibrium constant for ferrous heme is 10^{11} times better than carbon monoxide. This affinity is fine if the receptor for NO is a hemoprotein. However, the NO synthase would be unable to release NO as a product if this were the end of the story. The inherent instability of cytochrome b's and their requirement for thiols allows them to transfer the nitrosonium ion (NO^+) to thiolate (S^-), protecting the enzyme from product inhibition and generating a releasable NO. The reaction scheme can be summarized as follows:

$$NO^+ + RSH \leftrightarrow RSNO + H^+ \tag{14}$$

$$RSNO + RSH + Me^{+3} \leftrightarrow RSSR + NO^{\cdot} + Me^{+2} + H^+ \tag{15}$$

$$NO^+ + 2RSH \leftrightarrow RSSR + NO^{\cdot} + 2H^+ \tag{16}$$

where Me^{+2} or Me^{+3} is a transition metal ion.

This process can be catalyzed on a hemoprotein surface as follows:

$$Heme(Fe^{+2})NO^+ \cdots RS^- \leftrightarrow heme(Fe^{+3})RS^- + NO^{\cdot} \tag{17}$$

$$Heme(Fe^{+3})RS^- \leftrightarrow heme(Fe^{+2})RS^{\cdot} \tag{18}$$

$$Heme(Fe^{+2})RS^{\cdot} + RSH + NO^+ \leftrightarrow heme(Fe^{+2})NO^+ \cdots RS^- + RS^{\cdot} + H^+ \tag{19}$$

$$Heme(Fe^{+3}) + RS^{\cdot} + RSH \leftrightarrow heme(Fe^{+2}) + RSSR + H^+ \tag{20}$$

$$Heme(Fe^{+2}) + NO^+ + RSH \leftrightarrow heme(Fe^{+2})NO^+ \cdots RS^- + H^+ \tag{21}$$

Superoxide could replace RSH but would yield peroxynitrite (^-OONO), which would in turn yield hydroxyl radical and nitrite.[70,233] The hydroxyl radical is very short lived and very reactive. As noted for GHP, it is unlikely to escape the heme reactive site before reacting. The result would be self destruction of the heme. NO synthase is capable of acting as a weak NADPH oxidase that generates hydrogen peroxide, presumably from recombination of hydroxyl radicals; like GHP, however, the turnover is likely to be limited because of the self-destruction (see Chapter 2). The following series of reactions summarize this process:

$$O_2^- + NO^{\cdot} \rightarrow {}^-OONO \tag{22}$$

$$O_2^- + NO^+ \rightarrow {}^{\cdot}OONO \tag{23}$$

$$1e^- + H^+ + {}^-OONO \rightarrow NO_2^- + HO^{\cdot} \tag{24}$$

$$2e^- + H^+ + \cdot OONO \rightarrow NO_2^- + HO \cdot \qquad (25)$$

$$2HO \cdot \rightarrow H_2O_2 \qquad (26)$$

Therefore, as thiol levels decrease and oxygen levels increase, NOS will self-destruct, based on reactions (22)–(25). High thiol levels in the midst of limited oxygen maintains NOS activity.[78,451,452] The substrate inhibition of peroxidase, green hemoprotein, NADPH oxidase, and NO synthase by 3-amino-L-tyrosine supports the idea that NO synthase like that of these other enzymes contains protoporphyrin IX-type heme noncovalently bound in the enzymes' active site.[218] Furthermore, the poor peroxidative and hydroxylating activities of NO synthase suggest that it is a type-b cytochrome (see Chapter 2). Although it has not been observed, NO should inhibit NADPH oxidase activity by binding to the heme in place of oxygen, causing release of the heme as a NO-heme complex. That NO synthase is a cytochrome b was suggested in January 1991 and confirmed 28 July 1992.[453] Finally, like GHP (phosphoenolpyruvate decarboxylase ferroactivator), NOS reacts most dramatically (i.e., the NO it produces) with iron (or transition metal) nonheme sulfur proteins.[454,455] However, unlike GHP, which is positive on the target enzymic activity, NOS has a negative effect on its target enzymes.[233,449,456-460]

II. A SELF-CONTAINED REDOX CHAIN

A. STRUCTURE AND PROPERTIES

At first, it was suggested that the same cytochrome b component might serve NADPH oxidase (superoxide synthase) and NO synthase in leukocytes because of their similar inducers, NADPH oxidase activity and inhibition by 3-amino-L-tyrosine.[218] However, they have been shown to be distinct type-b flavocytochromes, although intimately connected in macrophage redox functions, perhaps, even with exchangeable hemes.[231,461]

The NOS family is divided into inducible NOS (NOSi) and noninducible, or constitutive, NOS (NOSc).[79,108,462-464] The inducible form can be subdivided further into enzymes regulated by calcium and those that are unregulated.[60] Fundamentally, the mechanism of NOS is not completely understood but its structural similarity both to cytochrome P450 reductase, in its high sequence homology at the flavin dinucleotide (FAD) and flavin mononucleotide (FMN) binding sites, and to cytochrome P450 itself, in the heme optical absorbance spectrum, suggests similarity in mechanism.[453,463,465] NOS has a broad maximum absolute light absorbance at 406-nm wavelength and a reduced carbon monoxide difference spectrum with the characteristic near 450-nm peak centered at 447 nm.[453,466] Furthermore, the activities of macrophage NOS and cerebellum NOS were inhibited by carbon monoxide (62–79%). Using the reduced pyridine hemochrome assay, NOS was shown to contain 3.3 to 6.5 nmol of protoporphyrin IX heme per milligram of protein. The theoretical

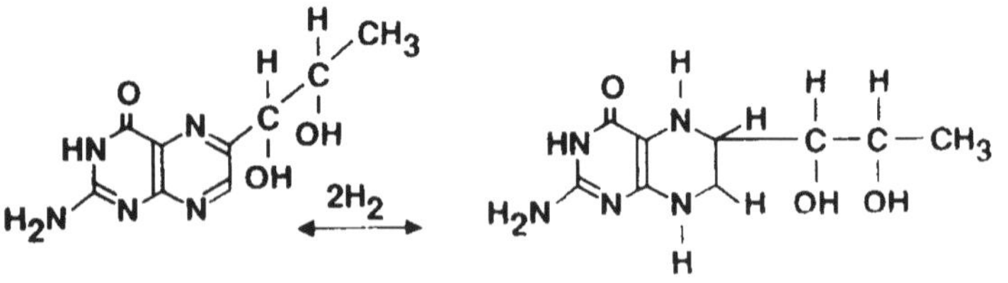

FIGURE 11.   The oxidized and reduced structures of L-erythro-biopterin.

content of the macrophage 130-kD protein is >7.69 nmol of heme per milligram of protein. Like other type-b cytochromes, NOS has the tendency to lose its heme.

Other common features of nitric oxide synthases are that they range in molecular ratio (M_r) from 125 to 160 kD, require oxygen, NADPH, 6(R)-tetrahydro-L-biopterin (Figure 11), bound FAD and FMN, and thiol.[452,467] The biopterin must be recycled by a specific reductase.[468] The thiol is usually glutathione.[452] The primary substrates are oxygen, NADPH, and L-arginine.[451] Arginine and oxygen are the ultimate sources of the NO.[451] The process of oxidation of arginine is a complex five-electron transfer.[451] Without thiol, the enzyme autoxidizes yielding hydrogen peroxide, with rapid inactivation.[70]

Like NADPH oxidase, there is an apparent oxidatively sensitive activity-essential thiol and a requirement for hydroxyl free radical in the process. Although the hydroxyl radical may be required in NOS activity, it must be constantly controlled to prevent self-destructive autoxidation of the heme, which is again typical of type-b cytochromes. The NOS has an unusual absorbance-spectral shoulder at the 450- to 475-nm region, which is similar to that of *Bacillus megaterium* fatty acid monoxygenase P450 BM-3.[453,469] This spectral property results from interaction between the flavin and heme.[453] The macrophage NOS has one tetrahydrobiopterin, FAD, FMN, and heme per 130-kD monomer. Tetrahydrobiopterin is typically involved with nonheme iron hydroxylations as in phenylalanine hydroxylase and in association with molybdenum in nitrogen oxide reductase of plants and bacteria (see Chapter 6).[453]

The inducible and constitutive nitric oxide synthases from liver and brain, respectively, have consensus calmodulin-binding sites.[60,463] The liver, or hep-NOS, a NOS inducible with IL-1, TNF, gamma-interferon, and LPS, has an 80% homology with NOSi from macrophages and a 50% homology with endothelial and brain constitutive NOS.[60] Therefore, it is actually a third intermediate form of NOS between inducible and constitutive nitric oxide synthases. This form of NOS is also found in aortic smooth muscle tissue.[60] The distinction between NOSi and NOSc based on calcium control has been further muddled by the fact that a constitutive membrane-bound calcium-controllable NOS has been found in macrophages.[470]

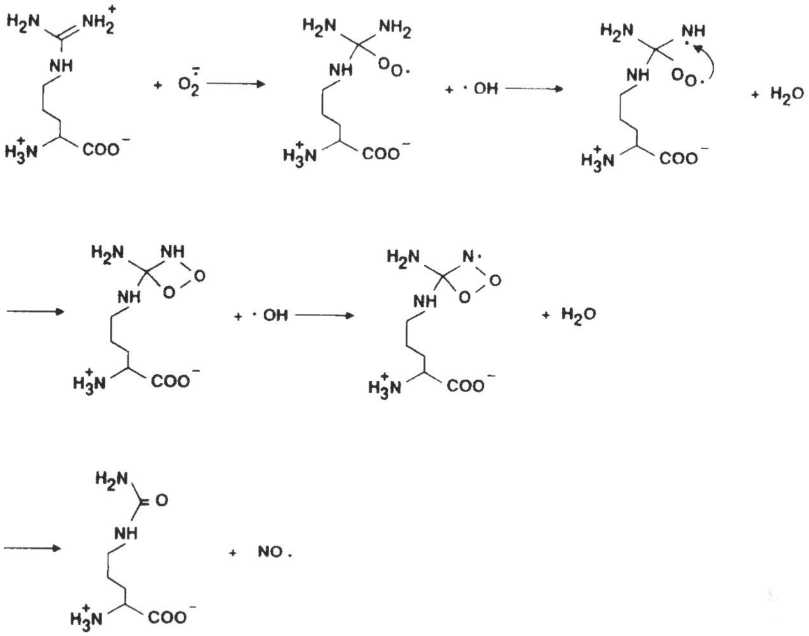

FIGURE 12. A hypothetical reaction mechanism for the oxidation of L-arginine by nitric oxide synthase.

B. ENZYMIC MECHANISM

The above structural considerations and similarities to other type-b cytochrome redox chains (nitrogen oxide reductases, green hemoproteins, NADPH oxidase, and cytochrome P450) suggest certain mechanisms of action. A hydroxylation mechanism like that of cytochrome P450 has been suggested, in which perferryl iron heme complex, $(FeO)^{+3}$, hydroxylates the quanidino nitrogen of arginine.[453] But beyond this step, the explanation for tetrahydrobiopterin participation and molecular oxygen as the only source of the oxygen NO is not forthcoming.[453] Also, the five-electron process has been difficult to explain by a cytochrome P450 mechanism.[453] An additional suggestion that a Fe(III) peroxide anion species is involved in the final step of the reaction series has been made.[453] This peroxide involvement includes a P450-like hydroxylation of arginine followed by a peroxide attack on N^G-hydroxyl-L-arginine. However, the fifth electron step to form NO and citrulline can only be awkwardly reached. The reduction of the iron in the heme has been proposed, for this final step.[453] Figure 12 shows a more elegant mechanism here for the first time.

The alternative scheme is a nucleophilic attack of the superoxide anion onto the guanidino nitrogen of arginine, resulting in a peroxy radical of arginine. The hydroxyl free radical then attacks the guanidino nitrogen (which has become a primary amine), resulting in a diradical, or triplet state, which recombines to a cycloperoxide. This cycloperoxide breaks down through the peroxidase activity of the hemoprotein of NOS, releasing hyponitrous acid, which then reduces the ferric heme to ferrous heme. The resulting NO strongly binds to the ferrous heme. The addition of molecular oxygen would cause autoxidation of the heme to yield ferric heme, releasing superoxide and NO. The superoxide then would re-enter the next turnover cycle of arginine oxidation.

This alternative process overcomes two biochemical problems: (1) the autoxidation of NO in the active site of NOS and (2) the high binding constant of the reduced heme iron of NOS for NO.[70] The conversions of hyponitrous acid to nitric oxide and protons (acid) and of the ferric to ferrous heme cause a dissociation of the heme from the coordinated histidine base of the protoporphyrin IX-type hemoprotein.[70] This opens up the possibility that nitric oxide-heme complex may diffuse out of NOS and directly activate target enzymes such as guanylate cyclase.[234,471] This action is likely to take place when oxygen levels drop very low so that the ferric heme cannot be regenerated and the free NO is unlikely to be released from the reduced heme iron. This would be one explanation of how inducible calcium-independent NOS might be turned off when thiol levels go so high that oxygen is depleted in the cell and surrounding medium.

In the alternative scheme, the biopterin could reduce the nonheme transition metal ion bound to the thiol. This reduced nonheme metal ion would, in turn, reduce the hydrogen peroxide, forming hydroxyl free radical. The superoxide could be produced by direct reduction of the heme and its oxidation by molecular oxygen. This reaction would protect the heme from autoxidative destruction by hydroxyl radical while making it accessible for the arginine hydroxylations. Also, the breaking of the cycloperoxy-N^G-arginine could be mediated by the peroxidase activity of the hemoprotein component of the NOS. This process suggests, that in the absence of arginine, NOS should have some superoxide dismutase- and catalase-like activity. This is illustrated by the following reaction:

$$2O_2^- + H_2O_2 + 2H^+ \rightarrow 2O_2 + 2H_2O \qquad (27)$$

GHP does demonstrate some superoxide dismutase and catalase activity (see Chapter 2). However, these activities have not been demonstrated in a futile cycle of NOS.

III. MECHANISMS OF CELL ACTIVATION AND DEACTIVATION

A. CYTOTOXICITY

Before addressing the mechanisms of nitric oxide synthase activation directly in the effector cells, the effects of nitric oxide and nitric oxide derivatives will be examined. Based on the overall premise of this book, nitric oxide can be considered to have existed in the cellular environment before the enzyme (redox chain) for its production evolved. Its effectiveness as a cytotoxic agent and its high affinity for hemoproteins, in general, make it a highly probable primordial driver for the association between nucleic acids and hemoproteins.

The primary targets of attack for NO are nonheme-iron-sulfur proteins and nucleic acid bases.[60,460,472] The NO forms complexes with the iron of the nonheme iron centers and/or sulfhydryl groups, resulting in their oxidation to disulfides or formation of nitrosothiols.[60,473,474] Key enzymes and redox chains sensitive to these actions are ribonucleotide reductase (a key enzyme in DNA synthesis), aconitase of the Kreb cycle of mitochondria, and complexes I and II of the mitochondrial redox chain (NADH:ubiquinone oxidoreductase and succinate:ubiquinone oxidoreductase, respectively).[60,459] Ribonucleotide reductase, which uses the coenzyme thioredoxin as its source of reducing equivalents for the conversion of ribonucleotide to deoxyribonucleotide, requires a limited amount of oxygen for activity.[475,476] If excessively reduced, its essential tyrosyl radical is eliminated, but can be regenerated on exposure to oxygen.[346] Therefore, the enzyme is particularly susceptible to free radicals like NO, which may not only remove the thiol co-factor, but also recombine with the tyrosyl radical to form nitrosotyrosine.

Furthermore, NO may interact with amino groups of deoxynucleotides on DNA to form diazobases that hydrolyze to the deaminated forms and nitrogen gas.[472] Diazocytidine is converted to deoxyuridine, methyl-2'-deoxycytidine is converted to deoxythymidine, deoxyguanosine is converted to 2'-deoxyxanthosine, and deoxyadenosine is converted to deoxyinosine by the diazotization and hydrolysis. Such changes when not lethal are at least potentially mutagenic, if not repaired. Thiocyanide may also play a role in stabilizing the NO so that it can have a longer lifetime and interactively reach to more distant nucleic acids.[472]

NO has also been shown to activate cytosolic adenosine diphosphate (ADP) ribosyltransferase which may be a positive or negative effect.[469,477] Poly(ADP)ribosylation has been shown to be involved in DNA repair, cell differentiation, and transformation of cells to malignant cells.[478] However, mono(ADP)-ribosyltransferases can be potent inhibitors of protein synthesis (ribosylation of the protein synthesis ribosomal elongation factor 2) and are responsible in part for the action of pertussis, cholera, diphtheria, and botulinum toxins.[477,479] A 39-kD protein of unknown function seems to be the primary target of NO-induced ADP ribosylation.[469,477]

The targets of toxicity for NO produced by NOSi in macrophages and neutrophils, regardless of mechanism, are tumor cells, parasitic fungi, protozoa, helminths, mycobacteria, viruses, and intracellular parasites in general.[60,228,471,480-485] The short life and reactivity of NO make it ineffective against extracellular parasites, unless there is very close contact between effector cells and targets.

B. SOLUBLE GUANYLATE CYCLASE AND ADP RIBOSYLATION

In limited quantities and when effects are mediated by derivatives of NO such as nitrosothiols or NO-iron-thiol complexes, where toxicity is decreased, NO becomes a messenger molecule.[234,474,486-493] Activation of the soluble guanylate cyclase, which requires a nitric oxide-heme complex for full activity, is considered the principal target for intracellular and intercellular communication with nitric oxide and its thiol-bound derivatives.[474,490] Although the increase in cyclic guanosine monophosphate (cGMP) has been associated with sequestration of cytosolic calcium, resulting in relaxation of smooth muscle and decreased platelet aggregation, the sequestration of calcium in fibroblasts is through a cGMP-independent mechanism.[494-496] Free heme with NO bound to it is as effective as NO in activating soluble guanylate cyclase.[60,408] Therefore, as alluded to earlier, releasing heme from a cytochrome b in the membrane or cytoplasm could yield a stable signal for activation of the guanylate cyclase. In general, cGMP seems to mediate relaxation, shutting off calcium-calmodulin-dependent constitutive NOS and hepatic NOS.[60]

When poly(ADP-ribose) polymerase, an enzyme involved in DNA repair, is inhibited, pancreatic islet cells are protected from macrophage NO-mediated cytotoxicity.[109,497] This result suggests a programmed-cell-death mechanism may be involved and initiated by NO when proper growth and repair signals are not present.[409,498-503] Such a response has been observed in mouse macrophages.[264] Cyclic ADP (cADP) seems to act on intracellular organelles which store calcium and have ryanodine-receptor calcium channels.[504] The result is an increase in the release of calcium similar to the release mediated by IP_3 release by phospholipase C associated with cellular plasma membrane receptors. Whereas intracellular calcium increases enhance NADPH oxidase activity through the protein kinase C phosphorylation cascade in macrophages, NOS (from brain at least) is inhibited as much as 77% by protein kinase C phosphorylations.[463] Although cAMP-dependent kinase phosphorylation of NOS has no effect on its activity, cAMP-dependent kinase of NADPH oxidase components activate the enzyme in macrophages.[463] Therefore, there is a balanced competition between activation of NADPH oxidase and NOS in cells that have both systems, such as macrophages and neutrophils.

C. NF-KAPPA-B AND HIV ACTIVATION

Activation of NF-kappa-B transcription factor as noted in Chapter 3 is a multipotential activator of genes and is associated with H-ras in its activation. Activation of NF-kappa-B leads to immunoglobulin synthesis in B lymphocytes, interleukin-2 production in T lymphocytes, and upregulation of latent HIV in infected cells.[349,353,354,505] This activation can also be nonspecific, as supported by the observations that ionizing radiation and ultraviolet light exposures activate NF-kappa-B through a nongenetic mechanism.[349]

Gamma radiation (700 R) stimulated the production of nitric oxide in liver, intestine, kidney, lung, brain, spleen, and heart.[350] These results indicate a link between radiation and nitric oxide production. Therefore, either the NOS is activated directly by radiation or, as suggested previously, the influx of calcium from membrane injury activates the NOS. Whereas nitric oxide is usually considered to be viricidal, in contrast, it upregulates HIV expression through activation of NF-kappa-B.[349] Radioresistance in cells is mediated through mutations in ras.[506] Latent HIV is also activated to be expressed by UV.[347] These facts taken together suggest that a cytochrome b, or NOS specifically, associated with ras, may be responsible for the programmed cell death of lymphocytes in response to both radiation and HIV. Further evidence for this concept is found in the observation that pulsed-HIV-infected human dendritic cells present virus and stimulate blast formation in target CD4+ T lymphocytes.[507,508] The stimulation process causes replication, virion production, and death of the T cells.[508,509] The examination of specific tissue cells will provide further evidence for the sensor-switch role of nitric oxide synthase.

IV. NITRIC OXIDE SYNTHASE IN CELL-TO-CELL COMMUNICATION

A. MACROPHAGES AND NEUTROPHILS

The nitric oxide synthase activity of macrophages was initially considered to be of one type, the inducible form.[461,464,467,480,510] However, recently, macrophages of mice were found to have a second type, a constitutive NOS.[470] Therefore, macrophages (at least in mice) provide one of the best examples of the dual function of NOS as an inducer of synthetic processes, or growth and differentiation, usually attributed to constitutive NOS, and as a killing agent, usually attributed to the inducible NOS. The difference between these two functions is very fine and may be considered merely to be a matter of concentration or rate of production of NO and of the stage of differentiation (or state) of the target cell.

To make matters more confusing, many of the same cytokines and immunostimulants that trigger superoxide and peroxide production also trigger NO and thiol production.[265,461] Elevated hydrogen peroxide production was triggered by PMA in mouse peritoneal macrophages after 48-hour treatment with gamma-interferon, TNF-alpha, TNF-beta, or granulocyte-macrophage colony stimulating factor (GM-CSF). During the culture period, gamma-interferon was the only cytokine to induce production of substantial NO and nitrite. In these cells, IL-1-beta, IL-2, IL-3, IL-4, interferon-alpha, interferon-beta, macrophage colony stimulating factor (CSF-M), and TGF-beta-1 were all negative for NOS induction. Incubation of gamma-interferon for 48 hours with LPS inhibited hydrogen peroxide production, but enhanced nitric oxide production. However, contrary to the possibility that NADPH oxidase and NOS might share common components, N^G-monomethylarginine inhibited NO production but not hydrogen peroxide production. TNF-alpha or -beta both enhanced gamma-interferon stimulation of NO production sixfold. Also, interferon-alpha or -beta in conjunction with LPS induced NO production in mouse macrophages.[460,461] Exfoliative toxins A and B of *Staphylococcus aureus*, but not enterotoxin A and B, nor toxic shock syndrome toxin 1, stimulated mouse LPS-responsive but not nonresponsive macrophages to synthesize NO.[511] Bacillus Calmette Guerin (BCG) and other mycobacteria also stimulate NO production in macrophages.[228] Although there is some commonality in stimulatory pathways for NADPH oxidase and NOS, they apparently diverge.

Also, human neutrophils and HL-60 (promyelocytic human leukemia cells) differentiated with dibutyl cAMP, produced NO when treated with N-formyl-L-methionyl-L-leucyl-L-phenylalanine (FMLP), platelet activating factor (PAF), or leukotriene B_4 (LTB$_4$).[510] Dibutyl cAMP differentiates HL-60 cells along the neutrophil pathway. Human monocytes and macrophages have been negative to date in the production of NO. This negative result was also demonstrated in HL-60 cells differentiated with $1,25(OH)_2$-vitamin D_3 along the monocytic cell line. These cells released superoxide when treated with FMLP, at a level fourfold lower than the dibutyl-cAMP-differentiated cells, but no NO. However, HL-60 dibutyl-cAMP-differentiated cells released NO and superoxide but at rates ten- to two-fold lower than normal human neutrophils. FMLP was a much better activator of both NO and superoxide release than PAF or LTB$_4$. In cases where both NO and superoxide were produced, superoxide dismutase potentiated NO production. In both HL-60 cells and human neutrophils, no baseline NO or superoxide production was noted.

The NO synthase of macrophages also demonstrates NADPH diaphorase activity and reduces nitrotetrazolium blue to insoluble formazan.[60,453,461,512-514] NO synthases are also capable of reducing cytochrome c by direct electron transfer.[515] Therefore, they create confusion with the superoxide assay that uses cytochrome c reduction. However, the NADPH diaphorase activity is uninhibited by SOD and should even be enhanced by it, based on the above observations.

Macrophage deactivating factor (MDF; a 13-kD protein) and TGF-beta-1, -2 and -3 inhibited NO synthesis.[412] The NOS inhibition was only during the induction phase (the first 3 hours). Once NO release started, the inhibitors were ineffective. The MDF was 10 times more effective against NADPH oxidase than NOS, and the inhibition by TGF and MDF occurred in macrophages with activated NADPH oxidase. TGF-beta also suppressed the release of the positive stimulator TNF-alpha by 60%, but increased its transcription in human monocytes. Fifty percent of NOS activity was inhibited by 7 nM MDF, 2 pM TGF-beta-1, 4 pM TGF-beta-2, and 8 pM TGF-beta-3. These cytokines partially blocked NOS induction in macrophages with gamma-interferon and TNF-alpha, but failed to block induction with gamma-interferon and LPS. MDF and TGF-beta-1 (a potent chemotactic agent) also interfered with the induction of Class II MHC antigens and transcription of IL-1 and platelet-derived growth factor (PDGF).

A clear-cut separation of the NADPH oxidase and NOS activities was made by observing these activities in the cell lines J774.16 and J774C3C.[231] The J774C3C cells, transformed with Abelson leukemia virus, do not produce superoxide like the parent J774 line, but both cell lines produce NO.

Even though MDF and TGF-beta-1 both affect NADPH oxidase and NOS, the mechanisms of these effects are unclear because these cytokines have no effect on particulate phagocytosis, glucose uptake, or NADH, NADPH, glutathione, and cytochrome b_{558} (of NADPH oxidase) levels.[412] The activation of protein kinase C is also unaffected by these cytokines. The NADPH oxidase and NOS activities in macrophages are generally considered to be upregulated by two different mechanisms: (1) *de novo* synthesis for NOS and (2) protein modification and assembly of the redox chain components for NADPH oxidase.

The macrophage NO synthase (Mac-NOS) has been purified and cloned from the mouse monocyte/macrophage transformed cell line RAW 264.7.[464,467] In general, NO synthases represent a unique, soluble P450-like cytochrome b that incorporates the reductase into the same protein.[453] The functional Mac-NOS (or NOSI) is a homodimer of 125- to 130-kD subunits.[78] The enzyme requires FAD, FMN, tetrahydrobiopterin, and thiol as co-factors and L-arginine, NADPH, and oxygen as substrates.[78,516] The K_m for arginine is 2–5 µM. The mouse Mac-NOS shows 52% overall homology with rat-brain NOS (NOSII) and a conserved region near the N terminus (residues 157 through 476 of Mac-NOS) that shows 71% homology with brain NOS.[463]

Both NOS I and NOS II have very similar binding sites for NADPH, FAD, and FMN in the C-terminal half of the enzymes.[463,464] The conserved region near the N terminus may be the L-arginine binding site or part of the enzymic active site. The Mac-NOS is more stable at 4°C or when turning over during electron transfer at 37°C than either cerebellar or neutrophilic NO synthases. NADPH-cytochrome P450-oxidoreductase is the only other known mammalian enzyme to contain both FAD- and FMN-like NO synthase.[453] Analysis of Mac-NOS has indicated that it contains two molecules of FAD and one of FMN per dimer.[467] This property could be a result of flavin loss or binding heterogeneity. The difference in relative mass observed on electrophoresis contributes evidence in support of the subunit heterogeneity hypothesis. Also, although NOS is a soluble cytosolic enzyme, it is found in both soluble and particulate fractions of macrophage cytosol.[470]

The biopterin required to support NOS activity in macrophages is either 7,8 dihydrobiopterin or 5,6,7,8-tetrahydrobiopterin (see Figure 11).[468] Half-maximal stimulation with biopterin is reached at 20–30 nM of reduced biopterin. Since NO generation has been obliterated by methotrexate and aminopterin, inhibitors of dihydrofolate reductase, only when dihydrobiopterin is used in place of tetrahydrobiopterin, the imminent co-factor must be the tetrahydroform. Dihydropteridine reductase (which is resistant to methotrexate and aminopterin) can replace dihydrofolate reductase in the cycle if the dihydroquinonoidal form of biopterin is substituted for the dihydro- form. Also, glutathione and, perhaps, other thiols can replace the enzymes as reductants to maintain biopterin cycling. Other enzymes that require tetrahydrobiopterin as a co-factor include phenylalanine hydroxylase, tyrosine hydroxylase, tryptophan hydroxylase, and alkylglycerol monooxygenase. The lack of biopterin has been suggested to be the reason human monocytes do not produce NO. In the presence of oxidized but not reduced co-factor, NOS is unstable when treated with 25 µM of L-N-methylarginine.[452] Based upon the type-b cytochrome nature of NOS, it is likely that the lack of an acceptable redox substrate leads to oxidative self-destruction of the enzyme. Further support for this hypothesis comes from the observations that even in the presence of pteridine reductases, NO synthesis, without thiols, diminished by 40–50% and that there was a delayed decline in rate, which was 70% reversible by the addition of thiols. Therefore, the thiols may provide protection against reversible and irreversible oxidative inactivation of the cytochrome.

The nitric oxide produced by macrophages and neutrophils attacks the nonheme-iron-sulfur protein (of mitochondrial electron transport and aconitase) and the thioredoxin-dependent ribonucleotide reductase, which is the rate-limiting enzyme of DNA synthesis, in target cells as well as those of the macrophages, themselves.[459]

Nitric oxide, either added exogenously or generated by stimulation with gamma-interferon and LPS, results in apoptosis, or programmed cell death in murine peritoneal macrophages.[264] This action was mimicked by glucose depletion, inhibition of the tricarboxylic-acid-cycle enzyme aconitase with fluorcitrate or inhibition of glycolysis with iodoacetate. However, effects on the mitochondrial nonheme-iron-sulfur proteins or electron transport chain, alone, did not yield apoptosis.

Programmed cell death or apoptosis, unlike necrosis, yields characteristic nuclear and cytoplasmic condensations and specific-endonuclease-mediated fragmentation of DNA into 180- to 200-bp oligonucleotides.[409,411,517-519] In necrosis, cellular organelles swell, membranes are disrupted early in the process, and the cell subsequently disintegrates. Although the DNA may be directly attacked by NO yielding deamination and mutation or indirectly by generating nitrosamines from primary amines that secondarily attack DNA, this kind of genetic damage is not the main driver of apoptosis.[472] NO cannot be considered the sole stimulus for apoptosis because ionizing radiation triggers it in lymphocytes and growth factor depletion triggers it in embryonic neurons and hemopoietic stem cells.[356,498,502]

However, the action of the gene product of bcl-2 is to act at the membrane level in an antioxidant pathway.[518] When mice are deficient in this product, they show hypopigmentation (melanocyte die-off and hypofunction), massive thymocyte apoptosis, polycystic kidney disease, and splenic lymphocytic apoptosis.[519] The splenic lymphocyte loss in conjunction with the thymocyte loss led to combined immunodeficiency disease. Bcl-2 is localized to areas of free radical generation such as mitochondria, endoplasmic reticulum, and nuclear membranes.[518] *N*-Acetylcysteine and glutathione peroxidase but not superoxide dismutase can counter apoptosis.[518] Hydrogen peroxide and menadione-induced autoxidation stimulate apoptosis in cells. Bcl-2 suppresses lipid peroxidation in cells, which may follow treatment with such pro-oxidants. Therefore, NO interaction with hydrogen peroxide or transition metals and peroxide, ionizing radiation or chemically induced hydroxyl free radical reactions can all be expected to induce apoptosis. But oncogene activation, in particular, c-myc, is associated with the process, and as noted earlier NF-kappa-B can be activated by oxidative processes, including NO interaction, which in turn activates expression of many genes including latent retroviruses such as HIV.[499]

High cell density in RAW 264.7 cells has been observed to turn on an oxidative process that leads to cell death when the cells are transferred to fresh medium.[218] A growth factor released from such stressed cells has been shown to prevent the death of other RAW 264.7 cells, if they are treated with the factor before being stressed. Antioxidants, such as 3-amino-L-tyrosine, protect cells from the oxidatively induced death process.[218,520] Concomitant with this increase in cell death and release of protective factor is an increase in spontaneous nitric oxide production by the RAW cells.[521]

Although NO has been seen as an immunosuppressive agent based on inhibition of mitogen-driven lymphocyte proliferation, it may activate lymphocyte differentiation and lymphokine production.[501,522-526] Evidence for the latter conclusion is found in the fact that TNF-alpha production goes up when human peripheral mononuclear cells are activated with NO.[522] This effect could have been the result of gamma-interferon production.[527] Further evidence for NO immune stimulation is found in the fact that expression of IL-2 in CD4+ T lymphocytes increases with NF-kappa-B expression, which follows NO stimulation.[353,354,458] Human peripheral blood mononuclear cells, monocytes and lymphocytes, have been shown to be activated by NO, as revealed by increased NF-kappa-B-DNA binding activity, enhanced rate of glucose transport, enhanced membrane-associated tyrosine phosphatase activity, and this activation, in turn, activated src family protein kinase p56[lck] activity in these cells.[522] A brain-specific isoform of this kinase, p56[fynB], has also been identified as a potential target for ultimate activation by NO. The NO induces membrane component clustering similar to that induced by antigen binding to lymphocyte cell membranes and actin polymerization induced by hydrogen peroxide and hemin.[60]

The resolution of these paradoxical effects of NO may be based not only on concentrations but also on feedback mechanisms. When mouse peritoneal macrophages were treated with pertussis toxin, which ADP-ribosylates the alpha subunit of certain types of G-proteins (i.e., G_o- and G_i-proteins), LPS-induced secretion of TNF-alpha was increased but NO production was inhibited under the same conditions.[479] A 41-kD protein was ADP-ribosylated in these macrophages. These activation pathways for both the human mononuclear cells and the mouse macrophages are cGMP-independent.

In 1992, the presence of both calcium-calmodulin-dependent constitutive NOS and inducible calcium-independent NOS was reported in RAW 264.7 mouse monocyte/macrophage cells.[470] Both types of enzymes were found in both soluble and particulate fractions of the cells. The constitutive NOS calcium sensitivity was demonstrated by stimulation of NO production with the calcium ionophore A23187. With increasing passage of the cells, calcium-dependent NOS decreased and overall NOS activity increased. The induction of NOS by LPS and gamma-interferon required sensitive thiols, since tosyl-lysyl chloromethylketone, normally a serine protease inhibitor that also affects hyperactive thiols, inhibited the activation of NOS, and the inhibition was partially reversible with glutathione or *N*-acetyl-L-cysteine. The glutathione had to be present at least 3 hours after exposure to the interferon and LPS to maximally reverse the inhibition.

In rat cerebellum and HL-60 cells, NO synthase activity leads to a similar ADP-ribosylation of a 39-kD protein (close enough to be the same protein as the mouse 41-kD protein).[477] These results provide strong circumstantial evidence for a shutdown-feedback mechanism stimulated by NO, acting through ADP-ribosylation. This ADP-ribosylation could also stimulate other gene expression, leading to retrovirus expression and programmed cell death. One killing mechanism of AIDS in human (HIV) and monkey (SIV) is through the stimulation of apoptosis.[409,503]

Another effect of NO is on cell adhesion. Cell adhesion (cytoskeletal effects) is inhibited by NO secretion even though the clustering and, therefore, adhesion may in fact stimulate NO production.[496]

In addition to negative-feedback mechanisms, such as the one involving cell adhesion, positive-feedback mechanisms also exist in macrophages for NOS induction or activation. Recent studies of the mouse mammary adenocarcinoma cell line EMT-6 showed that activated macrophage supernatants induced these cells to produce NO, which was in turn autocytotoxic.[456] The immunization of mice with BCG yielded peritoneal macrophages that produced the conditioned medium supernatants when treated with 50 ng/ml *E. coli* 0128:B12 LPS for 16 hours. This conditioned medium was applied to EMT-6 cells for 48 hours. These cells responded by producing NO, showing cytostasis and mitochondrial electron transport chain inhibition. The active ingredients of the conditioned medium were identified by antibody-inhibition and protein-purification techniques to be gamma-interferon and TNF-alpha. However, low doses of LPS, in the range used in the previous study, can make RAW 264.7 cells hyporesponsive in TNF production to subsequent doses of LPS.[528] When RAW cells were treated with 25 ng/ml of LPS for a minimum of 6 hours, they produced much less TNF when subsequently challenged with LPS. However, below 1 ng/ml, there was no suppression observed in these studies.

The result of increasing NOS activity with increasing passage number reflects other results of increased redox activity and NO production in RAW 264.7 cells with increased cell density and time in culture.[521] These observations suggest that nonspecific stimuli, such as physical stressors and high cell density, trigger induction of NOS and that constitutive NOS may be the initial sensor for the nonspecifically enhanced NOS activity. Solomon H. Snyder has said that nitric oxide synthase is one of the most sensitive enzymes in biology in keeping with its responsibility for fine-tuning the levels of a major physiologic mediator.[488] Therefore, in this case, it should be considered a prime candidate for the sensor.

The interactions of macrophage-generated NO with other cells are usually associated with killing or inactivation of the target. For instance, *Leishmania major*, the causative agent of Oriental sore, or cutaneous leishmaniasis, is killed by NO.[483] Viruses including ectromelia, vaccinia, and herpes simplex, are also inactivated, by NO produced by mouse macrophages (in RAW 264.7 cells and in normal ones). However, the action is not based simply on direct effects on the virus because NO had no effect on viral infectivity.[484] The cellular and viral replication machinery is vulnerable to NO. Besides the direct mutagenic potential effect of NO on nucleotide bases through deamination, the NO can inhibit the ribonucleotide reductases (of HSV1) and the cellular counterpart necessary for viral and cellular DNA replication, respectively.[459,472,484] Also, elevations in cGMP levels delay reactivation of HSV1 from its latent state (this elevation can be through guanylate cyclase activation with NO).[484,485] On the other hand, cAMP elevation (seen with macrophage activation) accelerates reactivation of HSV1.[484,530]

Lymphokines provide extra-macrophage regulation of NO production. IL-10, produced by $CD4^+$ lymphocytes of the Th-2 subset, not only inhibits synthesis of gamma-interferon by both T cells and NK cells, but also inhibits the synthesis of NOS by mouse macrophages.[524] Another lymphokine, IL-13, inhibits TNF production, generally downregulates macrophage LPS responses, but assists in upregulating gamma-interferon production.[529] This interleukin also inhibits HIV replication *in vitro* and immunity to tranplanted tumors. Also, NOS induction is directly involved in allograft rejection in transplantation, as one of the earliest events noted.[523] Even though NO has been shown to be generally immunosuppressive for lymphocyte proliferation, human lymphocytes can produce picomolar concentrations of NO that are essential as a messenger initiating lymphocyte responses, including DNA synthesis.[408,490,526,530]

B. ENDOTHELIAL AND SMOOTH MUSCLE CELLS

Unlike the intracellular NO-stimulated responses of macrophages, the actions of NO produced by endothelial cells and smooth muscle cells are mediated through soluble guanylate cyclase (150-kD protein).[60,490,492,494,495] This enzyme has been implicated as important in normal and abnormal cell growth. Lymphocytes in culture to which cGMP is added show stimulation of DNA synthesis and transformation to blast forms.[408] Cyclic AMP inhibits these responses. Cyclic GMP is also elevated in several tumors and in rapidly growing cell lines.[408] The activation of this enzyme is dependent on heme, NO and dithiols (extra- and intramolecular).[474] Various other nitrocompounds and oxidants may substitute for NO in increasing cGMP production. These include N-methyl-N'-nitro-N-nitrosoquanidine, sodium azide, hydroxylamine, nitroprusside, nitrite, 4-nitro-quinoline-1-oxide (a carcinogen), endoperoxides PGG_2 and PGH_2, fatty acid hydroperoxides, periodate, dehydroascorbate, peroxides, epoxides, and superoxide and superoxide dismutase (this action is inhibited by catalase but enhanced by nitrate reductase).[408] The nitrocompounds, unless they directly release NO, are converted to guanylate cyclase-activating material by oxygen.[408,441] Biochemicals that increase cGMP in tissues do so through oxygen and calcium-dependent mechanisms, apparently through a calcium-dependent NOS.[494,495] Among these biochemicals relevant to vascular tissues are cholinergic (such as carbamylcholine) and alpha-adrenergic stimuli, bradykinin, histamine (which also raises cAMP levels), and serotonin. The calcium dependence has been demonstrated by the increase of cGMP levels with the treatment of cells with A23187 and by the inhibition of this rise by calcium blockers such as tetracaine and verapamil.[495]

In vitro, cytotoxic levels of NO have been produced by isolated murine lung vascular endothelial cells treated with gamma-interferon and TNF.[531] This NO production was L-arginine dependent, and the NO production and tumor cell lysis (M5076 reticulum sarcoma cells) were inhibited by dexamethasone. These results indicate not only a physiologic role for NO production by endothelial cells, but also an antitumor and pathophysiologic one.

The action of NO on heme requires reduced heme for maximal binding and the subsequent action of NO on soluble guanylate cyclase.[494] Overt reduction of nitric oxide or its binding to thiols can inactivate it toward guanylate cyclase interaction. Therefore, high concentrations of dithiothreitol, cysteine, glutathione and ascorbate inhibit guanylate cyclase.[474,492,494] Also, antioxidants such as butylated hydroxyanisole (BHA), butylated hydroxytoluene (BHT), and retinoids inhibit activation of guanylate cyclase.[408] In addition, sulfhydryl-reactive compounds inactivate guanylate cyclase, including cadmium chloride, p-hydroxymercuribenzoate, N-ethylmaleimdie, arsenite, and iodoacetamide. The arsenite action is potentiated by British anti-Lewisite (BAL) or 2,3-dimercaprol.[408] Limited amounts of dithiothreitol or ascorbate (1 mM) plus hematin enhance NO activation of guanylate cyclase, indicating that limited reduction is stimulatory but excessive reduction is inhibitory to guanylate cyclase.[408]

Although hemoglobin has demonstrated inhibition of NO activation of guanylate cyclase, when the hemoglobin is presaturated with NO, it becomes an activator of guanylate cyclase activity.[449,492,532] This supports the notion that the receptor for intercellular NO interactions is a hemoprotein (containing a protoporphyrin IX-type heme noncovalently bound to the protein) that releases its heme after encountering NO (as the NO-heme complex).

There exist other stabilizers of NO, which normally has a half-life of 3–5 s, that can release NO on demand. One such major carrier is serum albumin, which carries the NO as an *S*-nitrosothiol adduct.[473] Human plasma was found to contain 7 μM *S*-nitrosothiols, of which 96% were *S*-nitrosoproteins, 82% of which were *S*-nitroso-serum albumin.[473] Free nitric oxide was only at about 3 nM. In rabbits administered 50 mg/ml of the NO synthase inhibitor N^G-monomethyl-L-arginine, *S*-nitrosothiols decreased by 40%. Of this decreased *S*-nitrosothiol, 95% were *S*-nitrosoproteins, of which 80% was *S*-nitroso-serum albumin. Associated with this change was a 22% increase in mean arterial blood pressure.

NO or the nitrosothiol-serum albumin is considered to be identical to endothelium-derived relaxing factor (EDRF).[473,492,533] EDRF not only acts as a vasodilator, but also prevents platelet adhesion. Both processes are mediated through activation of the soluble guanylate cyclase by NO or by the NO carried by thiols. Iron ions may also form dimeric-to-polymeric thiol complexes with NO and may be derived from nonheme iron sulfur proteins from redox chains attacked by NO.[474] Therefore, the NO can be released and reused to affect another redox chain after inhibiting the original redox chain.

The cGMP produced by the guanylate cyclase seems to be associated with a specific set of protein kinases that mediate the response.[408] The principal result of its activation in vascular smooth muscle and in platelets is the sequestration of calcium ions.[495] Therefore, the process that causes influx or intracellular release of calcium, which in turn triggers NOS activation, sets up the process for intracellular calcium sequestration. Studies have suggested that stimulation of phospholipase A_2 and inhibition of lysolecithin acyltransferase facilitates EDRF- and cGMP-mediated vascular smooth muscle relaxation.[495] Inhibitors of phospholipases antagonize NO- and cGMP-dependent relaxation, suggesting that it is mediated through a diacylglycerol- or inositol trisphosphate (IP$_3$)-mediated release of calcium.[494] The free fatty acids themselves may trigger the NOS activation. Nitrite, itself, may be a storage form of NO that upon acidification favors the equilibrium between HONO and NO to shift to NO, which diffuses away.[70,493,494]

The calcium activation of NADPH oxidase in leukocytes could be short-circuited by NO from the vascular endothelium, which would bring about sequestration of intracellular calcium. The function of NO in endothelial and surrounding cells could be changed from a physiologic one to a pathophysiologic one by changing the redox state of NO. The NO reacts with superoxide to yield peroxynitrite, which can decompose by single-electron reduction to nitrite and hydroxyl radical, the latter of which is cytotoxic.[458] The nitrosonium ion (NO$^+$) can form the nitrosothiol adducts that are most likely to have a physiologic role.[233,493] This hypothesis is supported by results using a preparation of NO

(1.2 μM) added to a ring of endothelium-free rabbit aorta culture, which showed marked relaxation for 15–18 min.[495] However, addition of superoxide dismutase to the preparation significantly lengthened the relaxation, and when xanthine and xanthine oxidase were added in place of SOD, the relaxation time was significantly shortened. Although acetylcholine vasodilation is in part mediated by NO, when endothelium-free rabbit central ear artery preparations were constricted by perivascular nerve stimulation, the acetylcholine treatment produced vasodilation by stimulating presynaptic muscarinic inhibitory receptors of adrenergic nerve terminals, which was not sensitive to *N*-monomethyl-L-arginine inhibition (L-NMMA).

Other physiologic effects of NO include rat penile erection mediated by NO from rat penile neurons innervating the corpora cavernosa and the neuronal plexuses in the advential layer of the penile arteries.[534] The arteries and cavernosa are dilated by NO released from the neurons. TNF-alpha, IL-6, and IL-2 treatment of isolated hamster papillary muscle led to decreased contractility, even after removal of the endocardial endothelium.[535] This effect was inhibited by L-NMMA. Therefore, the myocardium itself contains an NOS. These reversible effects occurred within 2–3 min and were maximum at 5 min, remaining constant for 20 min and completely reversing within 40 min. IL-1-alpha had no ionotropic effect. NO treatment of smooth muscle led to hyperpolarization.[495] A similar effect may occur in cardiac muscle. It appears that in certain tissues, perhaps endothelium and myocardium, the constitutive NOS is also turned on by cytokines as is the inducible NOS in macrophages.[536] Therefore, the shock effects of TNF, LPS, and side effects of IL-2 may be mediated not only through activation of NOS in macrophages and neutrophils, but also by release of cytokines that activate NO synthases directly in target tissues.

C. NERVOUS TISSUE

Even though NO production was first discovered in macrophages and endothelium, interest in the actions of NO in the nervous system has received the greatest attention, predating the discovery of NOS in the nervous system by 10 to 11 years.[234,457,486,487,537] As in the vasculature, guanylate cyclase is the principal mediator of NO functions in the nervous system.[60,234,325,489,491,538-540] Also, the importance of NO as a neurotransmitter provides a strong biochemical link between the nervous system and the immune system.

First, the neuronal effects of NO, then the properties of the neuronal NOS will be examined. The product is emphasized first because its effects and regulation are not totally dependent on the properties of the neuronal NOS, since influx of inflammatory cells and other resident nonneuronal cells such as vascular endothelium could potentially produce the same local effects as the neuronal NOS, but under very different control.[463,516] Furthermore, the nitrosothiols and nitrite reservoirs of NO mentioned previously represent a long-range potentially systemic involvement of all sources of NO in the organism.

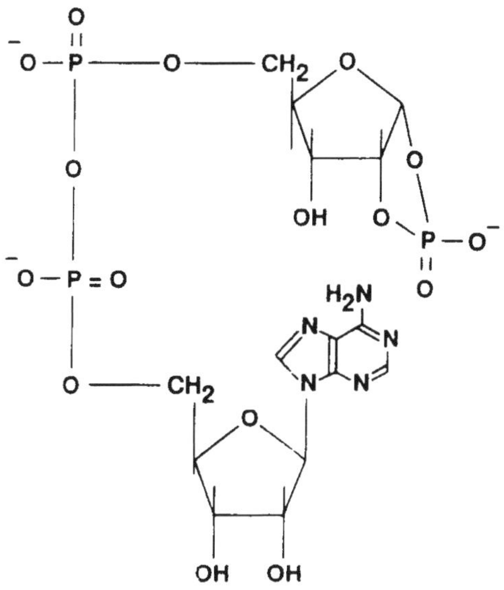

FIGURE 13. ADP-ribose-1″,2″-cyclic phosphate.

As discussed previously, guanylate cyclase, specifically its heme group, is considered the primary receptor for NO. However, cGMP-independent mechanisms, although less defined, should be considered in the action of NO in the nervous system. The alternative second messenger for this other pathway is cyclic adenosine diphosphate ribose (cADPR; see Figure 13).[469,477,504,541,542] It is formed as part of an RNA-splicing mechanism by an NAD-dependent phosphotransferase that catalyzes removal of the splice-junction 2′-phosphate.[542] This enzyme will remove an internal 2′-phosphate from a transfer RNA (tRNA) or from an oligonucleotide as small as a dimer, but not from a 5′-, 3′-, 2′-, or 2′, 3′-cyclic phosphate of an oligonucleotide as part of the process of forming ADP-ribose-1″, 2″ cyclic phosphate.[542] This cADPR is not just an incidental product of splicing but is an active mediator of intracellular calcium release from endoplasmic reticulum.[504] It is a more important releaser of this pool of calcium than IP$_3$ (1,4,5-trisphosphate inositol). All cell types so far examined have the machinery for its synthesis and degradation.[504]

cADPR mobilizes calcium by interacting with the ryanodine receptor.[504] This receptor was first described as the major calcium-release channel for sarcoplasmic reticulum of muscle. The receptor is similar to the IP$_3$ receptor in that they are both transmembrane proteins with long cytoplasmic domains and four identical subunits. Ryanodine receptors are involved with the self-potentiating property of calcium-induced calcium release (CICR). They are involved with oscillating calcium signals in sheets of cells, which can propagate as regenerative planar spiral waves.[504,541] Inhibition studies support cADPR's role in calcium release by ryanodine receptors. cADPR in the nanomolar range mimics caffeine, a ryanodine receptor activator, in inducing oscillations in calcium-dependent ion currents in rat dorsal root ganglion cells. cADPR sensitizes the ryanodine receptors to calcium, raising their level of excitability. This phenomenon, if carried to the extreme, can have pathological if not fatal consequences.

The disease associated with this pathology is malignant hyperthermia (MH). It is a runaway overheating of the body (in humans and pigs) associated with halothane anesthesia or succinylcholine treatment exacerbated by caffeine.[543] The disease is genetically heterogeneous, although many cases are associated with chromosome 19 and the ryanodine receptor locus (RYR₁).[543,544] This receptor defect causes hypersensitive calcium release in skeletal and cardiac muscle, resulting in the metabolic hyperthermia and clonic and tonic contractions of MH. Another genetic disease with a high propensity for MH is central core disease (CCD), which is an autosomal-dominant, nonprogressive myopathy, causing hypotonia and proximal muscle weakness in infants.[544] By 5–10 years of age, typical mild to moderate proximal muscle and facial muscle weakness is apparent. In pigs, MH is associated with a mutation at residue 615 of the ryanodine receptor. The arginine at this site is replaced by cysteine in the mutation. This residue is located at the cytoplasmic surface of the ryanodine receptor and may be an important regulatory site.

The gene of the plasma membrane cADPR cyclase of mammalian cells that makes cADPR has nucleotide sequence homology with that of the human lymphocyte surface antigen CD38.[542] This implies a receptor linkage to at least some cADPR cyclases. The implications for cADPR run even deeper, since the mobilization of calcium and its positive feedback could greatly enhance calcium-dependent NOS activation. NOS can, in turn, activate ADP-ribosyltransferase that ADP ribosylates ribosomes and depletes cADPR.[469] The effects are inhibition of gene splicing by cADPR depletion and inhibition of protein synthesis. The cGMP production stimulated by NO could then shut down the calcium-induced calcium release by cADPR depletion and activate calcium sequestration.

However, latent retrovirus expression, like that of HIV, requires inhibition of RNA splicing to express the structural genes of the virus (the regulatory genes are spliced RNA products).[545,546] This process would follow the shutting off of calcium release from the endoplasmic reticulum and the active export of calcium by calcium ATPase and the expulsion of calcium from the cell by the sodium/calcium plasma membrane antiporter.[547] This drop in intracellular calcium concentration should shut off the calcium-dependent NOS, allowing production of the viral structural mRNA without the threat of oxidation of the ribose units or deamination of nucleotide bases. ADP ribosylation of protein has also been associated with DNA repair and gene expression in mammalian cells.[478] Therefore, this ADPR-mediated process could be very important in the expression of HIV gene products and in the pathophysiology of AIDS in the central nervous system and in the immune system.

In 1977, a soluble component of brain synaptosomes, L-arginine was found to be an activating factor of soluble guanylate cyclase.[234,471,488] The L-arginine was found to be converted by rat brain synaptosomes to L-citrulline, with an accompanying increase in cGMP. This process was inhibited by L-*N*-monomethyl-arginine. In rat cerebellar cells, *N*-methyl-D-aspartate (NMDA), an excitatory amino acid, was shown to induce cGMP elevation.[234] The elevation of intracellular calcium ion associated with NMDA receptor stimulaton, therefore, triggered L-arginine metabolism leading to increased cGMP. Glutamate, a physiologic neurotransmitter, interacts with NMDA receptors and induces the highest level of cGMP in the Purkinje cells of the cerebellum.[538] These cells possess the highest levels of guanylate cyclase, GMP-dependent protein kinase and its protein substrate in the nervous system. Inhibition of this cGMP production with N^G-monomethyl-L-arginine established NO as a messenger molecule in the central nervous system.[489] Glycine, an allosteric modulator of NMDA receptors, induces accumulation of cGMP in the cerebellum *in vivo*.[489] Since NO can easily diffuse to surrounding cells, even in nervous tissue, cells that do not have NO synthase but do have guanylate cyclase could also be activated. Conditioned medium from glutamate-stimulated cerebellar granule cells added to primary glial cell cultures increased the cGMP levels in these cells.[489] Glutamate added directly to the glial cells had no effect. Furthermore, when culture supernatants from granule cells treated with glutamate and NO synthase inhibitor were added to glial cells, there was no increase in cGMP levels.

The "NO hypothesis" of intercellular messenger action in the central nervous system proposes that NO plays three important roles: (1) in the generation of synaptic plasticity and long-term potentiation; (2) in coupling blood flow to neuronal activity through NO's vasodilatory action; and (3) in mediating activity-dependent determination of neuronal connections during embryonic and neonatal development.[491] In these three functions, the property of NO diffusibility across membranes and among cells is predominant. This makes NO a very good candidate as a retrograde messenger, feeding back on the presynaptic neuron and signaling it to increase its output of neurotransmitter.[487,488] This process would link it to memory storage by strengthing the connection between the two neurons.[487] Long-term potentiation (LTP) is blocked by inhibitors of NO synthase while nitroprusside strengthens synapses like LTP.[486,537] Hemoglobin also blocks LTP, presumably by taking up the essential NO. The NO synthase inhibitors injected into rats *in vivo* made them lose their ability to learn spatial tasks. Kainate, another neuronal excitatory agonist, also stimulates cGMP production and induces NO release but through CNQX-sensitive non-NMDA receptors.[60] Glutamate inhibits this response to kainate. Granule cells of adults were shown to be less sensitive to NMDA receptor agonists than cells from young animals.[234,491] The efficacies of NMDA versus non-NMDA receptor agonists that stimulate cGMP increases changed with age. The possibility for more than one NO synthase exists in brain since L-NMMA inhibited NMDA and non-NMDA receptor-mediated cGMP increases *in vitro* in cerebellar tissue. However, L-N^G-nitroarginine (N-NARG) inhibited NMDA-induced cGMP increases biphasically in cerebellar tissue of immature animals, but monophasically in that of adult animals. These results show differential sensitivity, implying the presence of two enzymes in tissue of young animals. NMDA-receptor stimulation in the hippocampuses of young rats led to NO production. N-NARG demonstrated a monophasic intermediate K_i between that of the two values for young rat cerebellar NO production.

Besides activating guanylate cyclase in the nervous system, NO also stimulates the phosphorylation of the protein DARPP-32, a 32-kD phosphoprotein phosphorylated in response to activation of dopamine receptors.[60] After phosphorylation, DARPP-32 inhibits protein phosphatase I. *In vivo,* dopamine stimulates phosphorylation of DARPP-32 through AMP-dependent protein kinase. Application of sodium nitroprusside (which releases NO) to substantia nigra of rats leads to DARPP-32 phosphorylation, as well. Hemoglobin blocked the action of this nitroprusside. These levels of sodium nitroprusside also led to a fourfold increase in cGMP levels, but no change in cAMP levels. Membrane-transitting analogues of cGMP also led to phosphorylation of DARPP-32, while cAMP membrane-transiting analogues had no effect. Therefore, an NO/cGMP pathway for DARPP-32 phosphorylation exists in the nerve terminals of substantia nigra neurons.

Another mechanism for NO effects in the nervous system is through release of corticotropin-releasing factor (CRF) from hypothalamic neurons stimulated by IL-2, norepinephrine, or acetylcholine.[60,512] IL-2 and carbachol were additive but norepinephrine and IL-2 blocked the release induced by each agent separately or acted in tandem along the same stimulatory pathway. L-Arginine potentiated the effects of IL-2, while L-N^G-monomethyl-arginine prevented IL-2-induced or IL-2-and-carbachol-induced release. However, the inhibition did not effect norepinephrine-induced release of CRF. But when applied also with IL-2, the release was completely blocked. Atropine also blocked IL-2-stimulated CRF release. Three types of hypothalamic cells have been proposed as being involved in the CRF release induced by IL-2: (1) cholinergic neurons stimulated by IL-2 to release acetylcholine; (2) interneurons stimulated by acetylcholine to produce NO; and (3) CRF-containing neurons.[60] The NO stimulates cycloxygenase to produce PGE_2, which activates adenylate cyclase. The cAMP produced by the cyclase causes release of the CRF. A similar pathway could lead to activation of cAMP-dependent kinase, which phosphorylates DARPP-32. However, the carbachol-, IL-2-, and norepinephrine-induced results described above are somewhat inconsistent with this mechanism.[60]

Gaseous heme-binding messengers other than NO may exist in the nervous system. One such possibility is carbon monoxide.[537,548] This gas can be produced by the action of heme oxygenase. Inducible heme oxygenase-1, as discussed earlier, is a stress protein. However, the nervous system (the brain in particular) has a noninducible heme oxygenase-2 in highest concentration in the neurons of the olfactory epithelium and olfactory bulb. In the hippocampus, it is predominantly expressed in the pyramidal cell layer and dentate gyrus. In the cerebellum, it is evident in the granule and Purkinje cell layers and the pontine nucleus. The habenula, piriform cortex, tenia tecta, olfactory tubercle, and islands of Calleje also contain heme oxygenase-2. There is a strong parallel between the inducible and constitutive heme oxygenases and NO synthases. The rate-limiting enzyme in porphyrin synthesis, delta-aminolevulinate synthase is colocated with the heme oxygenase. Also, cytochrome P450 reductase, a source of electrons for heme oxygenase, is largely colocated with heme oxygenase in the brain. Furthermore, guanylate cyclase can be activated by CO. NOS inhibitors had no effect on cGMP levels in cultured olfactory neuron cultures, and these cells lacked NOS or NOS activity. Hemoglobin did, however, lead to depletion of cGMP in these cells, possibly, because of CO binding. When 50 mM of the heme oxygenase inhibitor zinc protoporphyrin IX (K_i = 3 mM) was applied to olfactory neuronal cultures, cGMP levels were lowered by 60%. The potent odorant 1-isobutyl-3-methoxypyrazine and several other odorants raised cGMP to a peak at 20 s to 1–2 min, depending on the odorant. This response was blocked by hemoglobin and zinc protoporphyrin IX, but not by nitroarginine. Both NO and CO have been shown to produce activity-dependent long-term synaptic enhancement (LTP) in hippocampal slices.[537]

Other parts of the nervous system show links between other cytokines and NO synthase. In the retina (bovine), synthesis of NO synthase is induced by gamma-interferon and LPS in the pigmented epithelial cells (RPE).[549] TGF-beta-1, which inhibits NOS induction in macrophages, slightly increases nitrite production in retinal pigmented epithelial cells stimulated by gamma-interferon and LPS. However, acidic and basic fibroblast growth factors (FGF) inhibit nitrite production induced by LPS and gamma-interferon. Proliferation of RPE was inhibited by NO production, and growth stimulation by FGF inhibited NO production. These results indicated an inverse correlation between NO production and proliferation. Laser retinal damage below thermal injury levels is related to oxidative events, and the initiation of this damage may be localized to the RPE.[550] NO is one potential source of this pathology. Combined with possible superoxide production from autoxidation of melanin and unsaturated fatty acids in the retina, NO production could lead to hydroxyl radical production and peroxide production. Degenerative and pathological effects have also been associated with NO in the rest of the central neurvous system (CNS).[232,500]

Circumstantial evidence has involved free radical oxidative damage with Parkinson's disease, Huntington's disease, amyotrophic lateral sclerosis, epilepsy, stroke, and hypoglycemia.[232,551] Furthermore, they have been associated with excitotoxic effects of glutamate when glutamate-gated calcium ion channels are excessively stimulated. In stroke models, such as ligation of the middle cerebral artery in mice, 1 mg of nitroarginine per kilogram of body weight administered in several doses postligation, provided 73% protection against stroke damage. In contrast, MK-801, the most effective NMDA antagonist, provided only 50–60% protection.

There is a 15-fold increase in mitochondrial DNA nucleotide oxidation with age. Hereditary mitochondrial DNA mutations and deletions are associated with neurodegeneration and stroke.[232] In the nervous system, the sources of reactive oxygen intermediates that may interact with NO include monoamine oxidase, tyrosine hydroxylase, L-amino oxidase, and autoxidation of ascorbate and catecholamines. As stated previously, NO may interact with superoxide to produce NO_2 and hydroxyl radical. Glutamate and other excitatory amino acids that lead to NO generation account for neurotransmitters of 40% of all synapses in mammalian central nervous systems.[325,552]

Three principal glutamate receptor types exist: (1) NMDA (*N*-methyl-D-aspartate) receptors; (2) AMPA (alpha-amino-3-hydroxy-5-methyl-4-isoxasole-proprionic acid) receptors; and (3) KA (kainic acid) receptors.[232,552] The NMDA receptor complex is composed of a transmembrane channel permeable to sodium, potassium, and calcium ions, with allosteric regulatory sites responsive to glycine, zinc ions, polyamines, and phencyclidine. A fourth glutamate receptor type is the metabotropic receptors, which mediate their action through G-proteins, much as cytokines and hormones do. These receptors modulate excitatory toxic effects of the other three receptor types. In neonatal mice, systemic administration of glutamate or aspartate leads to neuronal retina degeneration. When the arcuate nucleus of the hypothalamus was stimulated with pharmacological levels of injected excitatory amino acids, the neurotoxic lesions showed selective killing of neurons with cell bodies in regions of high glutamate concentrations, with sparing of axons from distant neurons and glial cells in these regions. Glutamate is removed from extracellular fluid by sodium ion-dependent high-affinity transporters of glutamate on glutamatergic neurons and glia.

Glutamate receptor-mediated neuronal degeneration is of two types: (1) acute and (2) delayed.[232] The influx of sodium during receptor overstimulation is followed by water that leads to swelling and rupture of neurons. Furthermore, the sodium/calcium ion exchanger, or antiporter, in these neurons, because of its bidirectional operation, counters the massive influx of sodium ions with a massive exchange for calcium ions.[517,547] This calcium, in turn, stimulates neurodegenerative processes.[517] Normally, they are prevented by a sodium/potassium-dependent ATPase, which pumps the sodium out, subsequently leading to calcium exchange to the outside as well. The calcium-dependent processes are responsible for delayed neurodegeneration, requiring several hours.[232,517] However, acute toxicity is not necessary for delayed toxicity. Increases in intracellular calcium ions may be mediated by voltage-dependent calcium ion channels or NMDA receptor-gated calcium channels. KA-induced neurotoxicity has shown calcium-dependent and -independent components.

Calcium activation of peptidases may convert xanthine dehydrogenase to xanthine oxidase, leading to superoxide production with ATP depletion.[232] However, calcium activation of NO synthase is another major source of neurotoxicity.[232] NOS inhibitors, reduced hemoglobin, depletion of arginine, and selective destruction of NOS-containing neurons all provide protection against NMDA-induced neurotoxicity in tissue culture. NOS neurons, themselves, are resistant to this toxicity.[232] However, mixed results have been achieved with *in vivo* systemic treatment with NOS inhibitors.[232] These equivocal results may stem from the level of superoxide or thiol present. Superoxide favors peroxynitrite formation and cell death, and thiols favor *S*-nitrosylation of the NMDA receptor thiol, which in turn downregulates the receptor and protects against neurotoxicity.

The killing of cortical neurons in primary cell culture by the HIV coat protein gp120, at very low doses, seems to follow the same pattern as the NMDA-induced neurotoxicity.[232] Glutamate receptor antagonists protect against this toxicity. Cell death is also prevented by NOS inhibitors, hemoglobin, and arginine depletion. However, conflicting data against NOS involvement in this neurotoxicity also exist. Superoxide dismutase also attenuates the gp120 neurotoxicity. These results may again suggest that NO neurotoxicity is conditional. The superoxide (which may result from complex I and II mitochondrial damage) may lead to hydroxyl radical production when it interacts with NO. Without the superoxide, the NO is far less damaging. Other toxins such as sodium azide demonstrate effects very similar to NO, suggesting that it is metabolized to NO in the nervous system.[514,553] Phenobarbitol at both anesthetic and subanesthetic doses provides the best protection against this lethality. This may be because the barbituates antagonize cGMP stimulatory effects and somehow deplete GTP, which is an allosteric inhibitor of glutamate dehydrogenase, the mitochondrial enzyme that metabolizes glutamate to the carboxylic acid alpha-ketoglutarate and ammonia, and therefore removes it as an excitatory amino acid. Oxygen treatment enhanced killing by sodium azide, suggesting oxidative conversion to NO.[553] The phenobarbitol may also directly compete with the oxidation of azide to NO through induction or interaction with a cytochrome P450 similar in structure to NOS, or phenobarbitol may bind to the type-b cytochrome of NOS. This hypothesis remains to be tested.

Inhibition of phospholipase A_2 or subsequent metabolism of arachidonic acid by lipoxygenase but not cyclooxygenases protected against glutamate neurotoxicity.[232] Also, induction of quinone reductase, which catalyzes two-electron reductions of quinones, decreases vulnerability to glutamate while inhibition potentiates the neurotoxicity. These effects are to be expected if a NOS that requires a semiquinone co-factor is responsible, at least in part, for the toxicity. Also, if hydroxyl radical is the imminent toxin in the neurotoxicity, then the eicosanoids or unsaturated fatty acid precursors would be expected to exacerbate the toxicity through autoxidative cycles that perpetuate hydroxyl radical production.

Apoptosis, or programmed cell death, which shares many characteristics with delayed excitotoxicity, is attributable to hydroxyl radicals that could result from NO and superoxide interaction and unsaturated fatty acid (UFA) hydroxyl radical propagation by autoxidation.[499,518,519] Glutamate-delayed excitotoxicity is itself delayed by inhibition of protein synthesis and DNA endonuclease activity.[232,411,499] Also, glutamate through an active carrier depletes neurons of cystine, particularly immature neurons.[232] This ultimately leads to glutathione depletion, which would again favor hydroxyl radical production from NO. Whereas astrocytes are resistant to this process, immature oligodendroglia are vulnerable.[500,554] Premature infants may develop periventricular leukomalacia via this process.

Parkinson's disease is a chronic, progressive disease of older people characterized by rigidity, involuntary tremor, bradykinesis, and degeneration of neuromelanin-containing neurons of the nigral dopaminergic type that innervate the caudate-pitamen.[232] Monoamine oxidase in these neurons increases with age. This enzyme catabolizes released dopamine and produces hydrogen peroxide as a consequence. In postmortem examinations glutathione peroxidase and glutathione concentrations were decreased in the substantia nigra of patients with Parkinson's disease. The pars reticula contains high concentrations of iron ions, which bind to neuromelanin and enhances hydroxyl radical production.

Chemicals that mimic Parkinson's disease in effect include 6-hydroxydopanine, amphetamines and their analogues, and 1-methyl-4-phenyl-1,2,3,6-tet·ahydropyridine (MPTP).[232] The latter chemical has created Parkinson's symptoms in humans and other primates. MPTP is metabolized by monoamine oxidase B of glia by being oxidatively converted to the cation 1-methyl-4-phenyl-pyridine (MPP$^+$). The cation accumulates in the mitochondria of dopaminergic neurons, selectively inhibiting complex I (nicotinamide adenine diphosphate: ubiquinone oxidoreductase). The inhibition has been noted in substantia nigra, platelets, muscle, and brain mitochondria in Parkinson's disease.

Transgenic mice made to overexpress copper/zinc superoxide dismutase are resistant to the pathology generated by MPTP. NMDA receptor antagonists protect against the neuronal degeneration of Parkinson's disease. This again implicates the NO/superoxide interaction as an important factor in this neurodegenerative disease.

Amyotrophic lateral sclerosis (ALS) is a middle-age degenerative disease of lower motor neurons of the spinal cord and upper motor neurons of the cerebral cortex that leads to paralysis and death.[232] In ALS patients, glutamate, aspartate, and *N*-acetylaspartylglutamate are at elevated concentrations in the cerebrospinal fluid. In contrast, the spinal cord and motor cortex per se show lowered concentrations of these neurotransmitters. Brittany spaniels that have a hereditary ALS show similar abnormalities in excitatory amino acid concentrations. As discussed earlier, autosomal dominant genetic ALS in humans shows defective genes for copper/zinc SOD.

Huntington's disease is an autosomal dominant genetic neurodegenerative disease with middle-age onset.[232] It shows signs of movement disturbances, psychiatric disorder, and progressive dementia. The striatum neurons show the most degeneration in this disease. The gamma-aminobutyric acid (GABA)/enkephalin-containing spiny neurons innervating the globus pallidus and the GABA/substance P neurons innervating the substantia nigra are the principal targets of degeneration. However, striatal medium and large aspiny and large aspiny neurons, internal capsule fibers and dopaminergic efferent neurons are spared. The cerebrospinal fluid levels of glutamate are elevated in Huntington's disease. NMDA receptor agonists mimic the neuronal specificity of Huntington's disease. Again, inhibition of complex I and II (whose electron transport is mediated by type-b cytochromes) and stimulation of NMDA receptors in neurons show the same neurodegenerative pattern as Huntington's disease. Susceptibility to inhibition of complex II increases with age.

Even in Alzheimer's disease, that has a primary molecular lesion unrelated to oxidative stress, unlike Parkinson's disease, ALS, and Huntington's disease, the pathology may be related to oxidative stress.[232] Amyloid A4 peptide, which increases in Alzheimer's disease, may increase vulnerability of neurons to glutamate excitotoxicity.

In an attempt to pursue NO function in the nervous system, biochemical properties of at least one NOS of the nervous system has been thoroughly examined.[79,463,466] Initially, with rat brain synaptosomal cytosolic preparations partially purified by ion exchange and gel filtration chromatography, properties of nervous system NOS were defined.[234] In the presence of NADPH, L-arginine induced synthesis of NO in these preparations. The synthesis was blocked by L-N^G-monomethyl-L-arginine at 200 μM. NOS was able to use L-arginyl-L-aspartate as a substrate in place of arginine. L-Arginine methyl ester was also a substrate but was less preferred. L-Homoarginine was 30-fold less potent as a substrate than L-arginine. D-Arginine, L-citrulline, ammonium chloride, L-alpha-amino-gamma-guanidino-butyrate and N-alpha-benzoyl-L-arginine ethyl ester were not brain NOS substrates. The K_m for L-arginine was found to be 8 μM; the brain normally has a level of 100 μM. Therefore, under normal conditions, the enzyme is saturated with this substrate. Brain NOS was inhibited by L-NMMA, N^G-nitro-L-arginine, and N-immino-ethyl-L-ornithine (L-NIO), but not by D-NMMA or L-canavanine. L-N^G-Aminoarginine and L-N^G-nitro-L-arginine are about 100-fold more potent than L-NMMA in blocking calcium-dependent NOS (as seen in endothelial cells).[555] In macrophages, N^G-amino-L-arginine and L-NMMA are about equal in inhibitory activity, whereas N^G-nitro-L-arginine is much less inhibitory.[555]

Kinetic studies showed that L-NMMA, N^G-nitro-L-arginine and L-NIO are competitive inhibitors of brain NOS with K_i's of 0.7, 0.4, and 1.2 μM, respectively, which is distinctly different from the aforementioned sensitivity in the endothelial calcium-dependent NOS.[79] The brain NOS only requires NADPH as an exogenous co-factor. However, EGTA buffers demonstrated the dependence of brain NOS on the presence of calcium ions. The enzyme was half-maximal in activity at 160 nM of calcium, with no activity demonstrable below 80 nM of free calcium ions. These levels of calcium stimulate NOS and inhibit soluble guanylate cyclase, providing a feedback control. The macrophage NOS requires exogenous tetrahydrobiopterin, which is not required by brain NOS.[468,516] L-Homoarginine is equal to L-arginine as an acceptable substrate for NOS in macrophages. However, L-homoarginine is a much less potent substrate not only for brain NOS, but also for endothelial and adrenal NOS. Also, as stated earlier, the NOSi of macrophages is not calcium dependent.[464]

In 1990, Bredt and Snyder reported the purification of rat cerebellar NOS to homogeneity.[79] They showed that the enzyme was calmodulin dependent and showed inhibition by the calmodulin antagonists trifluoperazine W-5 and W-13. They confirmed calcium dependency with EDTA inhibition and calcium enhancement of the NOS activity. The M_r of the enzyme was found to be about 150 kD by electrophoresis and 200 kD by gel filtration chomatography. The enzyme appeared to be a monomer. The apparent K_m for L-arginine was about 2 µM. The V_{max} was about 1 µmol/mg of protein per minute. The K_i for L-NMMA for the purified NOS was about 1.4 µM (twice that of the crude enzyme). The EC_{50} for calmodulin enhancement was 10 nM, and for calcium it was the same as the crude enzyme. The purified enzyme, typical of a type-b cytochrome, was unstable. When stored at 0°C, 50% of the NOS activity was lost in 2 hours, whereas crude enzyme required 2 days to lose this much activity at 0°C. The stability was increased by storage in 1 mg of bovine serum albumin per milliliter, containing 20% (v/v) of glycerol at −70°C. In this preparation, the NOS lost less than 50% of its activity in 7 days.

At least one type of brain NOS is also a neuronal NADPH diaphorase.[556] However, the cerebellar NOS appears to lack this activity.[465,556] NADPH diaphorase converts soluble nitro blue tetrazolium (NBT) to insoluble formazan. A physiologic function of this activity remains obscure. However, NADPH diaphorase is a selective marker for somatostatin and neuropeptide Y containing neurons and for the ascending cholinergic reticular system in the mesopontine tegmentum. NADPH-diaphorase-marked neurons are resistant to anoxia and excitotoxicity, and those in the striatum are selectively spared in Huntington's disease. NBT, also, competitively inhibits NOS activity, competing with L-arginine. The observed K_i was 11 µM. Also, where NADPH diaphorase activity was found, specific antiserum could be used to co-precipitate NOS and NADPH diaphorase activity. Cerebellar neurons that contain NOS were not stained with the labeled NADPH diaphorase antibody.

The rat cerebellar NOS has been cloned and expressed.[466] It was cloned as a cDNA with an open-reading frame of 4287 bases encoding a protein of 1429 amino acids with an M_r of 160 kD. The expression vector that was used had a cytomegalovirus promoter and was transfected into human kidney 293 cells. The protein expressed had a calmodulin-binding consensus sequence (amino acid positions 725–745), a cAMP-dependent protein kinase phosphorylation consensus sequence (at serine 473), no sequences for protein kinase C or calcium/calmodulin protein kinase phosphorylations, an NADPH-binding domain (amino acids 1204 to 1429), the contact points for ribose and adenine rings at amino acids 1245–1263 and 1343–1358, respectively. These sites are closely homologous with sites in cytochrome P450 reductase, sulfite reductase, and ferredoxin NADP+ reductase.[465]

Although brain NOS has no requirement for exogenous tetrahydrobiopterin, porcine brain NOS is stimulated by its addition. Consensus binding sites also exist in the rat cerebellar NOS for FMN and FAD.[465] Rat cytochrome P450 reductase has the greatest sequence homology with rat cerebellar NOS; the C-terminal halves of both proteins (641 amino acids) have 36% identity and 58% close homology.[465] The sulfite reductase is the next closest to NOS in sequence. It has 32% identity and 49% close homology in this region. The N-terminal half of this NOS has no sequence homology to any known protein examined.[465,466]

In 1992, Klatt et al. suggested that brain NOS was a soluble isoform of cytochrome P450 or cytochrome c reductase.[515] They used purified porcine cerebellar NOS to demonstrate calmodulin-dependent cytochrome c reductase activity. The reduction was superoxide dismutase and potassium cyanide resistant. The Km for cytochrome c was 34.1 μM. The V_{max} was 10.2 μmol/min/mg of protein; a rate 10- to 100-fold faster than NO formation (0.1 to 1.3 μ/mol/min/mg of protein). Binding of the porcine brain NOS to cytochrome c was demonstrated by affinity chromatography. When a reconstituted system of NOS and cytochrome P450 was tested for hydroxylation by demethylation of *N*-ethylmorphine, cytochrome P450 reductase activity was demonstrated by formation of formaldehyde. The Vmax for this process was 10 nmol/min/mg of protein. The reaction did not require calmodulin, was not inhibited by superoxide dismutase, catalase or both enzymes. Both NOS and cytochrome P450 were necessary for the reaction. NOS also reduced 2,6-dichlorophenoliodophenol (DCPIP) and NBT with a V_{max} of 3 μM/min/mg of protein and K_m's of 15.5 and 7.9 μM, respectively. The NBT reaction was calmodulin dependent, while the DCPIP reaction was not. These reactions inhibited NOS activity. The K_i's of cytochrome c, DCPIP, and NBT were 166, 41, and 7.3 μM, respectively. Tetrahydrobiopterin reversed these effects of DCPIP on the NOS activity of porcine brain NOS. These activities are what would be expected of a type-b cytochrome or green hemoprotein.

Studies of NOS from stably transfected 293S cells (selected by geneticin for transfection), transformed with both pCis-NOS and pRSV-neo plasmids, revealed that the NOS is stoichiometrically phosphorylated by protein kinase A, protein kinase C, and calcium/calmodulin-dependent protein kinase.[463] Protein kinase C activation *in situ* in the 293S cells (as by 12-*O*-tetradecanoyl-phorbol-13-acetate,

TPA) led to a 77% decline in NOS activity.[463] In cell homogenates, the decrease was 50%. The addition of the protein kinase inhibitor H-7, along with TPA, completely reversed the downregulation of NOS activity. One FAD and one FMN were found per molecule of NOS. It also appears that NOS is the only redox enzyme that requires calmodulin for activity.

In 1992, both inducible and constitutive NOS were found to be hemoproteins.[453,466] They both contain FAD- and FMN-binding sites, require NADPH as the electron source, require molecular oxygen as the source of the hydroxylating species of arginine, and require 6-(R)-tetrahydro-L-biopterin (H_4B) as an electron transfer mediator. In the case of the brain enzyme, the H_4B is tightly linked to the enzyme so that exogenous biopterin is not required. The macrophage NOSi was shown to contain a protoporphyrin IX-type heme by the pyridine hemochrome assay. This heme has a visible absorption peak (Soret) at 406-nm wavelength. This maximum implies that the iron of the enzyme is in low- and high-spin states. The broad apparent Soret is a combination of a high-spin peak at 389 nm and a low-spin peak at 416 nm. The prominent shoulder at 450–475 nm was attributed to FAD and FMN binding.[453,467] This property has been observed in P-450 BM-3 of *Bacillus megaterium*. The NOS should then contain one FMN and one FAD per 130-kD subunit. Dithionite-reduced macrophage NOS has a peak at 447 nm, much like cytochrome P450. Carbon monoxide and oxygen mixtures (80/20%) inhibited the NOS activity by 73–79%. Also, rat cerebellar NOS was inhibited by carbon monoxide, indicating the presence of a heme in the reaction center.[466] The presence of flavin nucleotide and heme on the same polypeptide indicates that NOS is the first water-soluble self-sufficient P450 cytochrome in eukaryotes (mammals).[453] Brain NOS has also been reported to contain nonheme iron. Both brain and macrophage NOS have been found to contain 2 mol of iron-protoporphyrin IX per mole of enzyme homodimer (4 hemes per NOS molecule). Specifically, the heme content was found to be 0.9 ± 0.2 and 1.2 ± 0.4 mol of heme per mole of monomer for brain (M_r = 150 kD) and macrophage NOS (M_r = 130 kD), respectively. The heme iron in resting NOS has been found to be ferric and five-coordinated with a cysteine thiolate as the proximal axial ligand. A stable bound flavin free radical (detected by ESR), or semiquinone, was found to be interacting magnetically with the ferric iron.

The purified rat brain NOS (cloned in 293 kidney cells) and macrophage NOSi (from RAW 264.7 cells) showed peak visible absorbances at 397 nm, a small shoulder at 420 nm, and broad but smaller bands at 538 and 650 nm, with flavin shoulders at 450–455 nm.[453,464,466] When dithionite reduced and treated with carbon monoxide, the peaks shifted to 448 nm (Soret) and 550 nm, again indicative of a P450-type cytochrome. As in cytochrome P450 reductase, the flavin semiquinone radical of NOS is stabilized and the terminal flavin is FMN. The interaction between the flavin and the heme, which accelerates the electron spin relaxation, is similar to that found in succinate dehydrogenase, where magnetic interaction occurs between nonheme iron-sulfur clusters and a flavin radical.[58] The relaxation in NOS is more characteristic of free iron, which shows cross-interaction with the heme iron and flavin radical.

Such a center of close interacting radicals, or electrons, is characteristic of a slow luminescence mechanism. The only other self-sufficient soluble cytochrome P450 is cytochrome P450 BM-3, as mentioned earlier, which also contains FAD, FMN, and heme in equimolar ratios and requires only oxygen and NADPH to catalyze hydroxylations of substrate. The heme-binding region of NOS is not, however, homologous to P450, as is the rest of the cytochrome P450 family.[453]

The homology of NOS with cytochrome P450 reductase suggests a function other than NO production. Heme oxygenase, which produces CO from heme oxidation, another potential neurotransmitter gas, requires electrons from cytochrome P450 reductase for its activity. NOS may be a soluble alternative source of these electrons. Furthermore, the decomposition of NOS may release heme that provides substrate for the heme oxygenase.

The slow luminescence mechanism that is essentially quenched in electron transport for the capture of useful biochemical energy may also be operable in the heme oxygenase. Zinc protoporhyrin IX, which is a highly potent and selective inhibitor of heme oxygenase, demonstrates slow fluorescence. Therefore, it may not be just a competitive inhibitor, competing against iron heme, but also an effective uncoupler of the electron-transfer process in heme oxygenase.

D. PANCREATIC CELLS (BETA CELLS)

The beta cells of the Islets of Langerhans of the pancreas are responsible for the release of insulin for the control of blood glucose levels.[109] These cells have been found to contain a type I NOS of 150 kD (NADPH- and calcium/calmodulin-dependent enzyme), which regulates the release of insulin.[109] The release of insulin is stimulated by L-arginine when D-glucose is present or by tolbutamide (a hypoglycemic-inducing drug).[109] L-NMMA inhibited the release of NO, increased intracellular cGMP levels in and increased release of insulin from beta cells.[109] The pancreatic A_2 cells (glucagon-containing cells) also contain NOS and this NOS is also a NADPH diaphorase.[109] The lower level of inhibition by nitro-L-arginine compared to L-NMMA and the lower molecular ratio of pancreatic NOS indicate that this NOS is different from that of the nervous system.[79,234,555] Also, this NOS can be activated by IL-1-beta, whereas that of macrophages is not, and the nervous tissue NOS seems to be primarily activated by the immunocytokine IL-2.[60,557]

Thus, the cross-talk between tissues that use NO as a second messenger or intercellular messenger is at least somewhat limited by the specificity of receptors linked to NOS activation. The L-arginine uptake in islets has a saturation of 15 mM, a 100-times-higher concentration than the normal plasma levels.[109] Therefore, plasma levels of arginine could regulate insulin release mediated by NO. This regulation of insulin release by L-arginine has been shown to be deficient in patients with non-insulin-dependent diabetes mellitus (type II), but not with insulin-dependent diabetes (type I). Diabetes mellitus has also been associated with decreased endothelial-dependent vascular relaxation and nonadrenergic, noncholinergic neurotransmission (a process mediated by NO).[109]

Studies of a cell-free system of islet microsomes has revealed that the principal second messenger in islet beta cells, triggering NOS activity and insulin release, is cADPR.[497,504] This conclusion is in opposition to the role of inositol-1,4,5-trisphosphate (IP_3) that has been accepted as the mediator previously. The former conclusion was based on the facts that heparin, an inhibitor of IP_3 binding to its receptor, did not block cADPR-induced calcium release; that cADPR desensitized islet cells to cADPR- and ryanodine-induced calcium release; that cADPR (1 μM) and calcium ion (0.2 μM) treatment of digitonin-permeabilized islet cells resulted in insulin release (this combination was not additive in effect); and that EGTA inhibited the cADPR effect. The maximum induction of insulin release induced by cADPR was achieved with 0.5 μM cADPR, and the Km of this activity was 0.09 μM.[504] This calcium release could also be triggered by high (20 mM), but not by low (2.8 nM) glucose levels. The maximal level of release of calcium by glucose was at 3 min post-treatment. The desensitization with cADPR did not affect IP_3-induced calcium release in these cells.

The diabetes-causing drugs streptozotocin and alloxan effect changes in the cADPR pathway indirectly.[504] The beta-cytotoxins cause DNA strand breaks, activating nuclear poly(ADPR) synthetase. This enzyme polymerizes ADPR of NAD^+ into poly(ADPR).[478] This competitive use of NAD^+ denies the ADP cyclase substrate, and, therefore, inhibits the cADPR-induction of insulin release.

Poly(ADPR) synthetase is a chromatin-bound enzyme in mammals.[478] The poly(ADPR) it produces is made up of ADPR units linked by glycosidic ribosyl (1″–2″) ribose bonds and ranges in size from a monomer to a large polymer of more than 50 ADPR units. Poly(ADPR) has been considered a regulator of histone function, nucleic acid metabolism, and gene expression. It may trigger repair processes triggered by nucleic acid damage. The enzyme has an absolute requirement for DNA and is stimulated by histones. Thiols and magnesium ions are also required for full enzymic activity. The enzyme has a pH optimum for activity at pH 9. The presence of histone has been observed to decrease the K_m for NAD^+ from 80 to 25 μM and double the V_{max} from 800 to 1560 nmol/min/mg of protein.

In the islet cells, the poly(ADPR) synthetase inhibitors nicotinamide and 3-aminobenzamide prevent the decrease in insulin production brought about by streptozotocin.[504] Streptozotocin-decreased intracellular calcium release in islet microsomes was reversed by nicotinamide (3 nM) and 3-amino-benzamide (0.4 mM). Whereas low concentrations of NO production should facilitate insulin release, high levels are likely to damage DNA, perhaps, activating the poly(ADPR) synthetase and stopping insulin release. If damage is severe, death of beta cells would lead to diabetes mellitus.

E. SKIN (KERATINOCYTES AND FIBROBLASTS)

Knowledge of NOS activity in skin is far more limited than in the four previously discussed tissues. However, in 1993, it was reported that primary human keratinocytes and a mouse keratinocyte cell line both demonstrated NOS activity when stimulated with gamma-interferon or TNF as well as 12-*O*-tetradecanoyl-phorbol-13-acetate.[307] L-NMMA inhibited this activity, and this NOS activity inhibited the growth of the keratinocytes. Epidermal growth factor was shown to be a potent inhibitor of keratinocyte NOS activity and to reverse the stimulation by gamma-interferon, TNF, and TPA.

Murine dermal fibroblasts have also been found to produce NO upon treatment with gamma-interferon, which was enhanced by TNF-alpha, IL-1, or LPS treatment (threefold).[558] The TNF-alpha, IL-1, or LPS were ineffective or minimally effective when given alone. For IL-1, 10 U/ml produced a maximal costimulatory effect. LPS produced a maximal costimulation at 100 ng/ml for 50 U/ml of gamma-interferon. At the same level of interferon, 500 U/ml of TNF-alpha produced a maximal effect. There was a steady increase in nitrite and nitrate production by the cells, in spite of cytotoxicity, over a 144-hour period.

The actions of gamma-interferon were traced, at least in part, to the production of tetrahydrobiopterin.[558] Gamma-interferon stimulates human cells *in vitro* to produce 100-fold the baseline level of guanosine triphosphate (GTP) cyclohydrolase I (EC 3.5.4.16) activity. This enzyme cleaves GTP into 7,8-dihydroneopterin triphosphate, leading to 7,8-dihydroneopterin and 5,6,7,8-tetrahydrobiopterin synthesis. This process occurs in human macrophages, fibroblasts, and a number of human tumor cell lines. Other enzymes critical to tetrahydrobiopterin synthesis are constitutive and not affected by gamma-interferon. These enzymes are 6-pyruvoyl tetrahydropterin synthase and sepiapterin reductase (EC 1.1.1.153). The GTP-cyclohydrolase I-induction was enhanced by TNF-alpha, LPS, or dexamethasone in human fibroblasts, human macrophages, and the monocytic cell line THP-1. Tetrahydrobiopterin is also an essential co-factor for phenylalanine4-, tyrosine3- and tryptophan 5-monooxygenases, demonstrating a mechanistic relationship between these enzymes and NOS. The oxidation of ether lipids is also mediated through tetrahydrobiopterin. The inhibition of GTP-cyclohydrolase I with 2,4-diamino-6-hydroxy-pyrimidine (5 mM) decreased nitrite formation and intracellular tetrahydrobiopterin levels by about 50%. The effect was reversed by treatment with sepiapterin, which replaces the *de novo* pathway by stimulating the salvage pathway. Methotrexate, which inhibits the salvage pathway, blocked the nitrite and nitrate production stimulated by sepiapterin. Also, inhibiting tetrahydrobiopterin synthesis provided protection against the cytotoxic effects of the cytokine and LPS. In conclusion, the data indicate that the skin NO synthases are related but are not identical to the macrophage NOSi (IL-1 insensitive).[558]

F. LIVER (KUPFFER CELLS AND HEPATOCYTES)

The NOS of liver, like the aforementioned enzymes of other tissues, has its own idiosyncrasies. Hepatic nitric oxide synthase (Hep-NOS) has been cloned and expressed in human 293 kidney cells.[60] Like macrophage NOS, this enzyme is inducible, but unlike MacNOS, it is regulated by calcium and calmodulin. This enzyme is also present in aorta smooth muscle.

Hepatocyte (HC) sensitivity to LPS cytotoxicity has been reported as dependent on coculture with Kupffer cells (KC) and on the presence of L-arginine.[321,333] By measuring damage to hepatocytes by release of aspartate aminotransferase, LPS effects were shown to be dependent on the presence of Kupffer cells. The ratio of Kupffer cells to hepatocytes had to be at least 7.5 to 1 to yield significant cytotoxicity (L-arginine had to be ≥ 1 μmol). Inhibition of protein synthesis also occurred and was evident at even lower cell ratios and arginine concentrations. The KC/HC cocultures show typical L-NMMA inhibition of NOS. However, the cocultures released more NO than did Kupffer cells alone in response to LPS, apparently because hepatocytes produce NO when exposed to conditioned medium from Kupffer cells (stimulated with gamma-interferon and LPS).[321,333,559]

In mice treated with *Corynebacterium parvum* followed by LPS, there was a marked increase in circulating nitrogen oxides.[324] However, L-arginine had no effect on the hepatic damage induced by these bacterial inducers.[324] L-NMMA had no effect in the absence of LPS but suppressed blood nitrite and nitrate levels in the presence of LPS. The inhibition increased the hepatic damage.[324] L-Arginine decreased the damage, partially reversing the L-NMMA effects. Such paradoxical outcomes may actually result from peroxynitrite formation. This unstable derivative is likely to breakdown rapidly into nitrogen dioxide and hydroxyl radical. The latter, although very reactive and, therefore, toxic, has a very short range due to its short life span in biological solutions. Furthermore, the numerous cytochromes and high level of thiols found in liver are likely to act as effective traps for the hydroxyl radical.

Some further confusion has arisen in the cytotoxicity of NO in the liver. Adamson and Billings have drawn the conclusion that NO plays no role in TNF and gamma-interferon hepatotoxicity because significant levels of NOS activity are not expressed until 4 hours after stimulation and because glutathione disulfide (GSSG) efflux and ATP depletion occur within the first 4 hours.[559] Also, 0.5 mM L-NMMA did not significantly decrease ATP depletion, GSSG loss, or lactate dehydrogenase (LDH) leakage. However, there was a substantial constitutive nitrite production (6 μM/24 hours) observed in control mouse hepatocytes. L-NMMA inhibited this production by only 66% (it could not completely inhibit production). Furthermore, in the presence of TNF and gamma-interferon, the same concentration of L-NMMA did not decrease nitrite production to or below the baseline constitutive level of activity (no lower than 7 μM). In addition, mannitol, benzoate, and iron chelators prevented ATP depletion and GSSG efflux induced by TNF-alpha. These inhibitors are all hydroxyl radical scavengers. However, when gamma-interferon was added, these inhibitors were only partially effective in protecting against GSSG loss and LDH leakage, but were ineffective in protecting against ATP depletion.

Also, depletion of glutathione by 1,3-bis(chloroethyl)-1-nitrosourea (BCNU), which is an inhibitor of glutathione reductase, resulted in greater loss of hepatocyte viability with TNF-alpha and gamma-interferon exposure than with TNF-alpha exposure alone.[559] In the hepatocyte preparations, L-arginine (1.15 mM) was present. It was reported that up to 5 mM L-arginine could not *completely* reverse L-NMMA inhibition in rat hepatocytes.[559] In another study, rat hepatocyte aconitase activity and mitochondrial complex I and II activities were inhibited by TNF-alpha by an apparent NOS-independent pathway.[324] However, these redox targets are precisely those anticipated to be most affected by NO. Although gamma-interferon also acts through nonoxidative mechanisms such as increasing the number of TNF-alpha receptors, the above evidence can be interpreted in another way. The constitutive NOS activity, which is apparently resistant to L-NMMA may yield NO that interacts with superoxide-producing metabolism to generate enough hydroxyl radical to further damage the mitochondria and other cytochrome systems. This process yields increased superoxide, lipid peroxidation, and hydrogen peroxide production. The induction of hep-NOS leads to more NO production. The NO and superoxide and/or hydrogen peroxide would interact to yield more hydroxyl radical. Also, the soluble cytochrome P450 activity of NOS (independent of the oxidation of L-arginine) could lead to cytochrome b or free-iron ion reduction and subsequent production of superoxide, hydrogen peroxide, and hydroxyl radical.

V. SUMMARY

What is abundantly clear in this review of nitric oxide synthase activity and NO effects in various tissues is that NOS does not simply exist in two distinct forms. Inducible and constitutive NO synthases may coexist in the same tissue and even the same cell. Intermediate forms that have some properties in common with both inducible and constitutive NOS exist. Furthermore, the inducers of NOS vary from tissue to tissue and cell type to cell type. Some inducers may act as inhibitors in other tissues and cells. In the face of the broad range of interactions that NO can participate in, these restrictions on its production and activation pathways provide some order among tissues and cell types that would otherwise be chaos due to the cross-talk due to NO diffusion. However, these restrictions do not completely preclude some very significant cross-talk among seemingly unrelated tissues, cell types, and cellular processes. The total result is complex profound pathophysiology in inflammatory and infectious diseases like AIDS. These properties also place NOS and its product, NO, in a central position in the sensing, switching, and intercommunication of all manner of cells in all manner of species.

Chapter 5

EXTRACELLULAR DISULFIDE REDUCTASE

CONTENTS

I. THE FEEDER CELL PHENOMENON

To date, extracellular disulfide reductase has not been shown to be a discrete enzyme or redox chain. Based on current perceptions of this cellular function, it could be the sum of transport and of intracellular reductase/dehydrogenase activities or a variety of separate transporters and enzymes acting in concert.[218,265,520,560-571] However, an alternate hypothesis can be made. This hypothesis is that the extracellular disulfide reductase is an alternate function of existing intramembrane components of another redox system and that the system can integrate with other cytoplasmic (soluble) redox components.[218] This switching of activities may be performed by a unique redox component that mediates the integration or a change in the internal/external molecular redox microenvironment.

Examination of these alternate hypotheses and candidates for the extracellular disulfide reductase (or its key switching component) begins with the "feeder cell" phenomenon. For many years, immunologists have observed that freshly isolated lymphoid tumors and normal lymphocytes have required feeder cells (usually macrophages or macrophage-like cells) for survival and growth in vitro.[572] This support was thought to be nothing more than supplying available thiols to the lymphoid cells when it was discovered that 2-mercaptoethanol (2-ME) could be substituted for the feeder cells *in vitro*.[572,573] Due to autoxidation, most of the cysteine in cell culture medium is in the form of cystine.[265,562]

Lymphocytes transport cystine with only 20% of the efficiency of transport of cysteine.[562,568,572] This efficiency is much less than it is for other cell types. The explanation for the ability to substitute a simple compound like 2-ME for feeder cells is that 2-ME forms mixed disulfides that are hydrophobic and, therefore, easily transported into the lymphocytes where they are reduced once again to thiols. Therefore, 2-ME was believed to act as a transport facilitator of cystine.

Further support for the enhanced-thiol transport mechanism comes from comparison of nonmacrophage cell lines to macrophage cell lines in their ability to produce extracellular thiols and support growth of the IL-2-dependent mouse cytotoxic T cell line CTLL-2.[265] In order of decreasing thiol production, the cell lines examined ranked as follows: Novikoff hepatocarcinoma > EL-4 thymoma > RAW 264.7 transformed mouse monocytes > P388D$_1$ transformed mouse macrophages. They showed the same ranking in support of CTLL-2 *in vitro* proliferation. Furthermore, agents that enhanced extracellular thiol production by RAW 264.7 cells (such as 0.02 to 2.0 µg/ml LPS for 24 hours) also enhanced support for CTLL-2 cells.[265] Also, agents that inhibited thiol production in RAW 264.7 cells, such as homocysteate, also inhibited CTLL-2 support. Cells that were nonresponsive to LPS stimulation for enhanced thiol production, such as P388D$_1$ cells, showed no enhanced support of CTLL-2 cells with LPS treatment. In contrast, high concentrations of LPS (2 µg/ml), in fact, inhibited the baseline support by about 66%. These observations were consistent with a simple mechanism of extracellular thiol growth support or facilitated transport of thiols.

However, paradoxes arose that suggested that this simple explanation of the feeder cell phenomenon was in error. First was the observation that 3×10^5 RAW 264.7 cells, which did not produce detectable levels of thiols, supported 79% of the maximal growth of CTLL-2 cells *in vitro*.[562] This result implied that supplying cysteine to the CTLL-2 cells for support of cell growth (primarily protein synthesis) was not a sufficient explanation for feeder cell support. This conclusion was based on the fact that even if lymphocytes transported only 20% of the available cystine compared to cysteine, the cysteine concentration maintained by 3×10^5 RAW 264.7 cells would need to be at least measurable to provide more cysteine than could be obtained by the lymphocytes themselves when the culture medium level is 0.2 mM. An alternative explanation was that the feeder cells were responding in a deliberate manner to the levels of oxidants in the medium and that this response was dynamically matched to the potential for oxidative damage.[562] Therefore, if this response was rapid and highly controlled (enzymically), no excess thiols would be produced. In the study from which these observations and conclusions were drawn, CTLL-2 cells constituted the only cell line that could not grow in the absence of 2-ME or a feeder cell population and produced no detectable thiols (even at high cell density).[562] Therefore, the adaptation of cell lines to culture, because of the abnormally high oxygen tensions *in vitro* when compared to *in vivo*, is accompanied by development of the capacity to produce extracellular thiols (as in the EL-4 lymphoid cells). Another line of evidence that supports the active production of extracellular thiols as a prerequisite for survival of cells in oxidatively damaging environments (as in culture) is that 2-ME, whether oxidized or reduced, supports CTLL-2 cell growth *in vitro*.[520,562]

Direct quantitative assessment of cystine uptake by lymphocytes supported by RAW 264.7 cells or 2-ME showed that, based on cysteine produced per unit of cells per 24-hour period, the cells were more effective in supporting CTLL-2 growth than an equivalent amount of 2-ME.[562] This effect may have resulted from a higher local concentration of thiols among the two cell types than with the 2-ME. Human neutrophils do not produce extracellular thiols when stimulated with endotoxin, particulate antigens, or phorbol esters nor do they reduce the disulfide of 2-ME.[520] Therefore, they lack an active extracellular disulfide reductase.

II. RESPONSES TO LIPOPOLYSACCHARIDE

Responses to LPS appear paradoxical based not only on the differences in results observed with different cell types (macrophages, polymorphonuclear leukocytes [PMNS], B lymphocytes, and hepatocytes), but also on the differences observed with cells in various metabolic states and with those treated with different lipopolysaccharides.[218,265,574] When the lipid A component of LPS, generated either by chemically breaking down 0111B4 *Escherichia coli* LPS or by using a J5 mutant incapable of adding polysaccharide to the lipid A, was added to human PMNS, superoxide generation and luminol chemiluminescence were stimulated.[575-577] At the same time, the lipid A destroyed the phagocytic, chemotactic and other metabolic activities of the PMNS. Toxic oxygen species were assumed to be responsible for these effects. But LPS can also stimulate cell growth and proliferation, as in B cells through a tyrosine kinase cascade.[576] It can also stimulate cells to support the growth of other cells as macrophages (at low cell density: 10^4 to 10^6 cells/per milliliter) are stimulated to do for T cells.[520] On the surface, mitogenic and antimitogenic responses, reactive oxygen intermediate generation, extracellular thiol production, NO production, and cytokine production (TNF, IL-1), stimulated by LPS, all seem to be mediated by individual pathways.[576,578-580] However, especially in induction of extracellular thiol production and cell growth responses, there appears to be an intimate connection.

For high-cell-density RAW 264.7 cell cultures (5.7×10^7 cells/per milliliter), even if the cells were held at this concentration only briefly (30–150 min), an inverse relationship was demonstrated between cell proliferation and extracellular thiol production.[218] In this case, the addition of LPS (011B4) at 100 pg/ml, during the brief period of high cell density, led to a decrease in extracellular thiol levels, but did not bring the level down to the low-density normalized concentration. Accompanying this decrease in thiol production per cell was an increase in cell proliferation. As noted before, increases in extracellular thiol production with 20 ng/ml to microgram-per-milliliter quantities of LPS led to proliferation of CTLL-2 cells supported by the thiol-producing RAW 264.7 cells.[520] However, exposure to as little as 0.5 ng/ml LPS for 4 days has been shown to inhibit RAW 264.7 cell proliferation by 50%.[218] Evidently, LPS triggers initial oxidative events that inhibit proliferation but can be countered by the simultaneous induction of reducing events (extracellular thiol production). High cell density seems to send chemical or physical messages similar to LPS and, like LPS, to trigger initial oxidative events, which lead to the overshooting reductive response. This extracellular thiol hysteritic response anticipates future oxidative insults and provides the appropriate protection. It also suggests that at least part of the redox machinery is shared between the oxygen-free-radical-generating redox chain and the extracellular-disulfide-reducing activity.[218]

III. RESPONSES TO ANTIOXIDANTS

If the increase in extracellular thiols were merely based on the transport of cysteine or peptides containing cysteine, then the addition of extracellular reductants or substrate that would drive the exchange (increasing the exchange rate) would yield a net increase in available extracellular thiols. The latter would occur if the oxidant levels were held constant and would be depleted if the substrates were unavailable or exhausted for reasons other than oxidation (i.e., protein synthesis). Such a transport system is found operable in IMR-90 human diploid fibroblasts in culture.[569]

These cells increase the concentration of sulfhydryl compounds (primarily cysteine) in the culture medium, and they convert medium cystine into cysteine.[569] The serum albumin from the fetal calf serum supplementation of medium (10%) has almost no free sulfhydryl groups in the unconditioned medium. It has been proposed that a cysteine/serum albumin sulfhydryl-disulfide reaction is responsible for the apparent increase in serum albumin sulfhydryls following culture of IMR-90 cells in the medium. Again, this is considered a response to the toxic levels of oxygen (18%) used during ordinary cell culture. In support of this notion, when oxygen levels in cultures were dropped to 1%, WI-38 human diploid cells gave better yields (at lower inoculum sizes of 10^3 cells per milliliter).

In 1988, Bannai and Ishii showed the operation of an "extracellular" cysteine/cystine redox cycle that involved three amino acid transport systems and an intracellular disulfide (cystine) reductase in IMR-90 human diploid fibroblasts.[570] For optimal growth of the cells, at least 2.5 mM glutamine was required. This requirement is based on a series of transporters. First, glutamine is removed from the culture medium by the System ASC. This amino acid transporter is sodium dependent (cotransported) and is especially reactive with amino acids with short or polar side chains. Sodium depletion of the cell culture medium leads to an 88–95% inhibition of the rate of transport of this system (varying with glutamine concentration). Second, glutaminase converts intracellular glutamine to glutamate. Extracellular cystine is exchanged as an anionic amino acid for the intracellular glutamate by the System X_c^-. The intracellular cystine is reduced by a reductase, and what is not used intracellularly is transported out by a gradient-driven transport system.

Glutamine transport by the System ASC is inhibited by L-alanine (81%), L-serine (80%), L-asparagine (65%), L-cysteine (86%, but not appreciably inhibited by cystine), L-leucine (49%), L-methionine (62%), and L-threonine (72%).[570] The K_m for glutamine transport is 0.25 mM with a V_{max} of 50 nmol/min/mg of cell protein. The K_i for cysteine inhibition of glutamine transport is 0.14 mM (close to the K_m for cysteine uptake). The Systems A, L, or N of amino acid transport were not operable in this exchange process. In summary, the System ASC takes up glutamine, which is deaminated, the intracellular glutamate is exchanged for cystine, the cystine is reduced by an intracellular disulfide reductase, and the cysteine is used to make proteins or glutathione, or it is transported out by a gradient-driven transporter. Only if the cysteine concentration rises too high in the medium does the System ASC stop taking up glutamine. Lowering the oxygen tension in the medium may effect this result. A uniform pool of 21 mM glutamate was maintained in the fibroblasts to drive the cystine/cysteine cycle, when the extracellular glutamate concentration was 0.2 mM, initially, and 0.5 mM after 1 day of culture. Glutamate is a competitive uptake inhibitor of the System X_c^- for cysteine.

A variety of other cell types use the System ASC for glutamine uptake.[570] These include rabbit reticulocytes, Chinese hamster ovary cells, rat mesenteric lymphocytes, and P388 murine leukemia cells (macrophage-like). Release of cysteine is mediated by the System ASC. Therefore, the loss of intracellular cysteine and the uptake of cysteine by this system is limited by the intra- and extracellular glutamine concentration, respectively. In RAW 264.7 macrophage cultures, the exchange of cystine for glutamate appears to operate since cystine declines and glutamate increases in the extracellular medium with days in culture.[581] Consistent with the fibroblast cystine/cysteine cycle involving these transporters, the alanine concentration in the medium rises significantly with time in cultures of RAW 264.7 cells, which may indicate transport of alanine out of cells because of the lower glutamine content due to the conversion to glutamate.

Mercaptoethanol enhances growth of lymphocytes *in vitro* that have limited cystine transport, as stated earlier. However, nonthiol reductants, or antioxidants, may or may not support lymphocyte growth and may or may not enhance cystine uptake. The two activities are not necessarily coupled as they are with 2-ME. Reduced ascorbate enhanced cystine uptake (30%) and cell viability, but did not enhance CTLL-2 lymphocyte net growth.[520] Levamisole and vitamin E increased cystine uptake (20–40%) and enhanced net growth in CTLL-2 cells (50%–fourfold).[520] The antioxidant 3-amino-L-tyrosine (3AT) inhibited cystine uptake by 30–60% and enhanced net growth by 1%–2.33-fold at 100 to 800 μM, respectively.[520] At 2 mM or greater, 3AT accelerated CTLL-2 cell death. At the pO_2 of 43 mm as opposed to the usual pO_2 of 147 mm in cell culture, 3AT and vitamin E showed no enhancement of growth of CTLL-2 cells over controls.[520] Mercaptoethanol, however, showed a 25–35% decrease in its enhanced growth support at pO_2 of 43 mm vs. pO_2 of 147 mm. Therefore, the nonthiol antioxidants operated wholly through an antioxidant mechanism of protection, and the 2-ME operated not only through this mechanism but also enhanced protein synthesis, a nonoxidative mechanism.[520]

In RAW 264.7 macrophages, which were stressed by high cell density and producing high levels of extracellular thiols, 3AT (0.2 mM) treatment of cells with 30-min pulses of 3AT resulted in about a 60% decrease in thiol production and a 30–40% increase in cell number over controls in 24 hours.[218] When this 3AT pulse included 100 pg/ml of LPS, the cell number increased 150% and the thiol production decreased only 50%.[218] LPS by itself decreased the thiol production 70% and increased cell number by 75%.[218] Therefore, the antioxidant activity of the 3AT inhibited the oxidation stimulated by the LPS and synergistically enhanced cell growth. Substituting oxidized 3AT for the reduced 3AT resulted in effects that were the same as those achieved with LPS treatment alone.[218] Therefore, the 3AT interfered with oxidation of the thiol by LPS-induced oxidants in RAW 264.7 cells, but in the process was not able to interact with the thiol-generating process as effectively.

These results suggest that 3AT must enter the cell and become oxidized in order to interact with the thiol-generating mechanism. It also suggests that both the oxidizing and reducing pathways induced by LPS are induced initially by the same pathway, which later diverges.[218] What has not been settled by these observations is whether the disulfide reductase per se is soluble, cytoplasmic, and coupled to the cystine/cysteine transporters or it is a transhydrogenase transporting hydride from internal thiols to external disulfides across the plasma cell membrane and then uptaking extracellular cysteine with a transporter. To complicate matters further, NOS is spontaneously increased in RAW 264.7 cells by high cell density.[521] Also, 3AT very effectively inhibits the oxidative burst of mouse peritoneal macrophages;[117] 2 to 3 μM of 3AT inhibit the burst by 90%. Therefore, several oxidative systems that require at least some thiol may be inhibited by 3AT and enhanced by high cell density.

There is evidence that, in mouse spleen lymphocytes, LPS and other mitogens upregulate the ASC transporter system.[565] This would lead to increased intracellular concentrations of glutamate and sodium. The glutamate would enhance the System X_c^-, therefore, increasing input of cystine for reduction. But the sodium ion concentrated in the cells could activate the sodium/calcium antiporter, leading to activation of other metabolic processes (such as constitutive NOS).

In Chapter 4, the control of NO cytotoxicity and its physiologic functions by thiols were discussed. Consequently, the NOS and disulfide reductase systems could be linked at this point not only in the immune system but also in other systems, such as the nervous system and pancreatic beta cells. In previous studies, the intracellular reduced glutathione (GSH) content of splenic lymphocytes was directly dependent on cysteine transport, and the increase in GSH following mitogenic stimulation positively correlated with increased cysteine transport.[568] However, lymphocytes lack the System X_c^- transporter. Therefore, the culture medium glutamine, serine, and threonine (all at 2.5 mM) were likely to inhibit the uptake of cysteine (by 88, 93, and 91%, respectively). Extracellular glutamine actually facilitates transport out of intracellular cysteine.[570] Activated lymphocytes have also demonstrated increased mixed disulfide (2ME-cysteine) transport.[568] This effect was accompanied by increased leucine transport because the mixed disulfide transporter also transports this amino acid.[403,582,583] This increased leucine transport also facilitates protein synthesis.

The question arises from these data as to how a macrophage feeder layer for lymphocytes can overcome the inhibition of cysteine uptake in lymphocytes by just releasing cysteine. Perhaps, in the close contact between the feeder layer and the supported cells, the glutamine and these other amino acids are depleted and the effective concentration of cysteine released is very high. Again, this inhibition of the System ASC under culture and physiologic conditions suggests the presence of an intramembrane thiol/disulfide transhydrogenase as the extracellular disulfide reductase.

In cells that form sheets and gap junctional connections, such as Chinese hamster V79 (743R) cells and normal human diploid fibroblasts (AG4322), reduced glutathione can be shared through metabolic cooperation, or actual transport from cell to cell.[567] This mechanism explains the cell-density dependence of cell population GSH content in these cells. PMA treatment blocks the equilibration of GSH content among sheet cells in a high-density population by disrupting cell-to-cell communication. Whether metabolic cooperation plays a role in the feeder cell phenomenon remains to be shown.

IV. RESPONSES TO NONSPECIFIC STRESSORS

The question arises as to whether the extracellular disulfide reductase, besides being sensitive to mitogens and antioxidants, can respond to physical stressors and microenvironmental changes. If it can, then it is linked to a true cell-level sensor. The X_c^- system, in cells that possess it, may still be a likely candidate for this sensor because as the extracellular cystine levels rise from oxidation and the intracellular glutamate pools are maintained at a high level by the System ASC and glutamine uptake, cystine uptake and reduction should both increase. The increased concentration of intracellular cysteine would increase *its* exchange for *extracellular* glutamine by the System ASC. The depletion of intracellular cysteine could be prevented by upregulation of glutathione synthesis.[560]

Glutathione synthesis is dependent on two ATP-requiring enzymes, gamma-glutamylcysteine synthase and glutathione synthetase.[560] GSH itself is not easily transported out of cells and would, therefore, accumulate cysteine intracellularly in its structure.[566] However, the glutathione disulfide dimer is readily excreted.[566] Again, without a steady-state supply of cystine or mixed disulfide-containing cysteinyl residues that could be easily taken up, oxidation and glutathione synthesis would rapidly deplete recyclable cysteine in the microenvironment. A thiol/disulfide transhydrogenase would solve this problem.

The most obvious stressor that should logically upregulate extracellular thiol production is oxidation. As stated earlier, lowering pO_2 from 147 to 43 mm increased the growth response of CTLL-2 cells and eliminated the growth stimulation by 3AT and vitamin E. Studies with 2-ME at the lower oxygen tension showed that 25–45% of the 2-ME effect on CTLL-2 cells was accounted for by antioxidant effects, with the remaining 55–75% stimulation of growth being associated with protein synthesis.[520] The improvement in growth by 3AT over control cultures was 28–46%, comparable to the antioxidant protection of 2-ME.[520] But these effects are small when compared to the thiol and growth responses in RAW 264.7 cells to high-cell-density stress.[218]

When RAW 264.7 mouse monocyte/macrophage cells were subject to brief high-cell-density stress (5.7×10^7 cells per milliliter for 30 min), and subsequently, cultured at a moderate cell density (1×10^6 cells per milliliter), the normalized thiol production increased sevenfold, while the cell number in 24 hours decreased sixfold.[218] These results suggest that some form of cell-to-cell communication participates in a self-amplifying redox process. However, the most important conclusions to be drawn from these results are that these cells are profoundly sensitive to microenvironmental changes brought about by their own presence and "remember" the change through synthetic and metabolic processes.

In this same study, interactions, or "cross-talk", between other stimuli and high cell density were noted.[218] The sensitivity and self-amplifying properties of the redox sensor, as measured through cell number and normalized extracellular thiol production, could be adjusted by these interactions. To generate combined effects, RAW cells were held at 5.7×10^7 cells per milliliter for 30 min while being exposed to 2450 MHz microwave radiation at 100 mW/g at 37°C and treated with 0.1, 0.5, or 1 ng/ml LPS. After 24 hours of culture (seeding at 1×10^6 cells per milliliter), the cell number significantly increased over incubator controls for cells treated with 0.5 or 1 ng/ml LPS and microwave radiation. This increase was 40 to 100% above controls. When sham exposures were performed under the same conditions but without microwave radiation, a significant increase in cell number over incubator controls was noted only for cultures treated with 1 ng/ml LPS. This increase was smaller at 30% above controls. The results indicated that the sham manipulation of the cells was responsible for 30–75% of the microwave response.[218] Therefore, the redox state of the macrophage determines its responsiveness to minor stimuli. The responses may, in fact, be paradoxical.

In previous studies, paradoxical responses of RAW macrophages have been noted in those treated with 3AT and 0.1 ng/ml LPS.[218] When compared to incubator controls, cells heated to 40°C for 30 min showed an increase in normalized extracellular thiol production, and cells heated to 42°C for 30 min showed a decrease in thiols. However, when microwave radiation was used to heat the cells to 42°C, the decrease in thiol production was significantly inhibited. The 40°C treatments had no effect on cell proliferation. However, heating to 42°C for 30 min (with 3AT and 0.1 ng/ml LPS) with microwave radiation showed significant decreases in cell number after 24 hours of incubation when compared to control or sham exposed (but heated) cells. When 0.5 or 1 ng/ml LPS was used with 3AT and 42°C treatments (microwave or sham heating), the cell number in 24 hours decreased by 45 and 58%, respectively, when compared to incubator controls. Correspondingly, thiol production decreased 57 and 58% (not normalized), respectively, for the two different concentrations of LPS. Therefore, it appears that LPS and 3AT sensitize the RAW cells to microwave radiation effects and hyperthermia. The microwave effects appear to be the same as hyperthermia effects but shifted to lower temperatures.

There are two key intracellular thiols important in thermotolerance and resistance to oxidative stress and radiation;[584-588] they are glutathione and thioredoxin.[475,476,560] However, they are also important in mediating some oxidative killing and cell-metabolism-modulating functions.[589] Treatment of Chinese hamster V79 cells with DL-buthionine-*S,R*-sulfoximine (BSO) or dimethylmaleate (DEM), which depletes glutathione by two different mechanisms, sensitizes the cells to heat killing.[588] The first compound inhibits the key glutathione synthesis enzyme gamma-glutamylcysteine synthetase and the second compound binds directly to reduced glutathione.[560] They even thermosensitize the cells to the thermotolerizing temperature of 42.5°C.[588] In these cells, reduced glutathione levels were decreased to 22% of control values by BSO and to 38 and 49% of control values by DEM. However, DEM was more effective than BSO at thermosensitization, suggesting that other thiols or mechanisms were operable in the thermosenitization process than just decreases in GSH. The GSH was also shown to rise with 1-hour exposures to 6% ethanol which also led to thermotolerance. Also, 5 hours post-treatment with ethanol, cells became more resistant to adriamycin, which is a cytotoxic agent that can generate anionic radical hydroquinones and oxygen radicals. Extracellular thiols such as cysteamine or cysteine promote heat sensitivity. The amino groups of these and other compounds have been associated with thermosensitization. The GSH levels in thermotolerance peak 1 hour after exposure, and thermotolerance itself requires 2–5 hours of heating. Therefore, the induction of thermotolerance by GSH is likely to be an indirect synthetic effect.

When GSH production in hepatocytes was inhibited with bischloroethyl nitrosourea (BCNU), a glutathione reductase inhibitor, the inducible NOS activity was inhibited.[590] This result suggests that the induction of a synthetic process by GSH may be through NO, which in turn activates transcription factors, such as NF-kappa-B. The additional fact that thioredoxin is essential in the killing of target tumor cells (such as HeLa cells) by gamma-interferon suggests a role for intracellular thiols in NO signaling because gamma-interferon is a powerful inducer of NOS.[589] However, thiol mediation of free radical formation is not limited to NO, especially at alkaline pH's (7.8 to 10.2) and in the presence of transition metal ions.[577,591] GSH, thiourea, cysteamine, dithiothreitol, 2-ME, and ethyl mercaptan, at low concentrations, can serve in this capacity to generate abundant superoxide and hydroxyl free radical through autoxidation.[577] However, the actions of thioredoxin are not limited to these simple mechanisms.

V. THIOREDOXIN

Thioredoxin (a dithiol-protein of $M_r = 12$ kD) may, in fact, manifest its most important effects through specific interactions.[475] For instance, it regulates the glucocorticoid steroid receptor, fos/jun heterodimers and has been suggested to regulate phospholipase C type I.[476,592] Thioredoxin can also be secreted.[475] This extracellular form demonstrates many of the activities of IL-1.[564]

Thioredoxin is a modulator of key enzymic activities in prokaryotes, plants, and animal cells.[475,476] Fructose-1,6-bisphosphatase, a key enzyme in the reductive pentose phosphate cycle of carbon dioxide assimilation in chloroplasts, is activated by thioredoxin. In plants, this activation is light-regulated by light-driven chlorophyll-reduction of ferredoxin, which in turn reduces thioredoxin through a ferredoxin-thioredoxin reductase.[476] Another enzyme of the reductase pentose phosphate cycle of chloroplasts, NADP-glyceraldehyde-3-phosphate dehydrogenase, is also light-activated by this reductive mechanism. Thioredoxin is widespread in cells being found in chloroplasts, algae, photosynthetic bacteria, roots, seeds, etiolated plant shoots, fermentative bacteria, aerobic bacteria, and animal cells.[475]

In animal cells and bacteria, it has been most often associated with ribonucleotide reductase, the rate-limiting enzyme in DNA synthesis. This enzyme converts ribonucleotides to deoxyribonucleotides and requires reduction of enzymic disulfides and the formation of a tyrosyl radical for the activity.[476]

Thioredoxin is also a bacterial-supplied subunit of T_7 phage-induced DNA polymerase in *E. coli*.[475] The viral coded subunit of this enzyme is the gene 5 protein (M_r = 84 kD). These subunits form a 1:1 heterodimer. In *E. coli*, the thioredoxin may be in the form of a phosphothiol, indicating that it may play a role in phosphotransferase reactions. However, it can be replaced in NADPH-dependent ribonucleotide reductase with glutaredoxin (M_r = 12 kD), which utilizes GSH as the hydrogen donor.

Sedoheptulose-1,7-bisphosphatase is another enzyme of the reductive pentose phosphate cycle of chloroplasts that is activated by thioredoxin.[476] Other plant enzymes activated by thioredoxin are phosphoribulokinase, NADP-malate dehydrogenase, and phenylalanine ammonia lyase. Thioredoxin also activates the perhaps nonphysiologic, ATPase activity of heat-solubilized preparations of chloroplast coupling factor (CF_1). In blue-green algae, 2'-phosphoadenosine 5'-phosphosulfate sulfatransferase (PAPSST), involved in sulfate reduction, is activated by reduced thioredoxin. The dark inactivation of the plant enzymes fructose 1,6-bisphosphatase, phosphoribulokinase, and phenylalanine ammonia lyase requires a soluble oxidant such as oxidized glutathione or dehydroascorbate. NADP-malate dehydrogenase is inactivated by a membrane-bound oxidant in the dark. The mechanism of dark inactivation of NADP-glyceraldehyde-3-phosphate dehydrogenase is unknown, but could be through ADP ribosylation, as is the inactivation of animal glyceraldehyde-3-phosphate dehydrogenase, which is associated with a redox mechanism.

Activation and deactivation of these plant enzymes are slow in respect to their catalytic activities. Therefore, hysteresis exists in their *in vivo* activation and deactivation. Light-induced changes in pH and magnesium ion concentrations in chloroplasts facilitate the thioredoxin regulation of these enzymes.[476] In light, stromal pH is shifted from 7 to 8, and magnesium ion concentration is increased from 1 to 3 mM. Fructose-1,6-bisphosphatase and sedoheptulose-1,7-bisphosphatase are particularly sensitive to these combined effectors.

Thioredoxin functions are so highly evolved in plants that different forms exist. Spinach chloroplasts contain thioredoxin f and m and thioredoxin c functions outside the chloroplasts in the plant cytoplasm.[475] Thioredoxin f activates the enzymes of the chloroplasts previously noted except for NADP-malate dehydrogenase, which is specifically activated by thioredoxin m.[476] The functions of thioredoxin c are unknown.

The reduction of thioredoxin in animal cells is catalyzed by a NADPH-dependent FAD-containing reductase.[475,476] Both the reductase and the thioredoxin contain catalytically essential disulfides. Thioredoxin interacts with ribonucleotide reductase through a Ping-Pong mechanism with a K_m value between 2 and 5 μM. Both thioredoxin and ribonucleotide reductase are highly conserved in primary polypeptide structure among all the species in which they are contained.

The thioredoxin in conjunction with its reductase forms an intracellular protein disulfide reductase.[475] Thioredoxin is in high concentration in both

Escherichia coli and calf liver (15 μM). Erythrocytes also contain thioredoxin, even though they lack ribonucleotide reductase and DNA synthesis. Because of the high content of GSH in eukaryotic cells (1–10 mM) most intracellular proteins are in the thiol form, and a few may be part of mixed disulfides with glutathione. Fibrin may be an extracellular protein reduced by thioredoxin (28 disulfides are found in fibrin or fibrinogen). Reoxidation of the reduced fibrin leads to tough, highly cross-linked fibrin, which may be important to coagulation and clot formation.

The E_0' value of *E. coli* thioredoxin is –0.26 V, which means it can reduce GSSG (GSH has an $E_0' = -0.25$ V) and cystine (cysteine has an $E_0' = -0.22$ V).[475] The E_0' value for protein disulfides varies from –0.20 to –0.40 V, meaning that some with more negative values cannot be reduced by thioredoxin. However, protein disulfides are still better substrates for thioredoxin than L-cystine or glutathione.[476] Insulin has been shown to be the best substrate with a K_m of 11 μM. Thioredoxin inactivates the disulfide-containing proteases FIX and FX, which are coagulation factors.

Calf liver and rat Novikoff hepatocarcinoma thioredoxins are different from the bacterial (*E. coli*) thioredoxin in that they can be inactivated by milder oxidation.[475] The mammalian thioredoxin reductase also has a broader substrate spectrum than the bacterial enzyme in that it can use thioredoxins of different species and the artificial thiol reagent 5,5'-dithiobis-(2-nitrobenzoic acid) (DTNB). Human thioredoxin has been shown to be an autocrine associated with IL-2 receptor upregulation after human T cell lymphotrophic virus (HTLV-I) transformation of these cells.[564,593,594] The next section deals with the receptor-linked activities of extracellular disulfide reductase.

VI. REDOX ACTIVITY OF LIGANDS AND RECEPTORS

The extracellular disulfide reductase, if it exists as a discrete enzyme system or redox chain that demonstrates hysteresis, should have both immediate effects and generate long-term synthetic changes. A novel location and means of transferring extracellular disulfide reductase has been suggested by the thioredoxin interaction with the IL-2 receptor and the disulfide isomerase activity of follitropin (follicle stimulating hormone; FSH) and lutropin (luteinizing hormone; LH).[564] These polypeptides have 60 to 300 times the disulfide isomerase activity of thioredoxin, respectively, in reactivating reduced and disulfide mismatched ribonuclease. Figure 14 shows the similarities in the active sites of protein disulfide isomerase, thioredoxin, LH, and FSH.[564] In humans, the LH contains an arginine substituted for histidine (unlike the ovine, bovine, and porcine structure shown). Even though LH and FSH have 11 and 12 cysteine residues, respectively, and 21 cystine residues in each of the approximately 35-kD complexes, the beta subunit of the heterodimer (these

TD:	- W - C_{32} - G - P - C_{35} - K -		
PDI:	- W - C	- G - H - C	- K -
LH-BETA:	- H - C	- G - P - C	- R -
FSH-BETA:	- H - C	- G - K - C	- D -
SS:	- A - G	- C - K - N	- F - F - W - K - T - F - T - S - C -

FIGURE 14. Disulfide reductase/isomerase active-site amino acid sequences in various polypeptides and a potential inhibitor. TD = thioredoxin; PDI = protein disulfide isomerase; LH-BETA = luteinizing hormone beta; FSH-BETA = follicle stimulating hormone beta; SS = somatstatin, which inhibits the release of insulin and glucagon stimulated by somatotropin; W = tryptophan, C = cysteine, G = glycine, P = proline, K = lysine, H = histidine, R = arginine, D = aspartic acid, A = alanine, N = asparagine, F = phenylalanine, T = threonine, and S = serine.

hormones contain a common alpha subunit) contains the reductase/disulfide isomerase activity. Furthermore, oxidation-reduction interactions between insulin, prolactin, prostaglandins, and catecholamines and their receptors have also been suggested.[564] Other active peptides that have vicinyl cysteine residues with basic amino acids that may stabilize a thiolate anion that would arise in redox reactions are nerve growth factor (NGF), snake neurotoxins I and II, and non-neurotoxic basic protein, cytotoxin II from *Naja naja* venom.[404,595] Therefore, active thiols in receptor interactions may be a common motif for polypeptide hormones and toxins. The redox changes in receptors induced by disulfide (isomerase) reductase activity of polypeptide hormones or cytokines, if they have an effect on transduction, should affect phosphorylation.

Both FSH and LH transduce their activities through cAMP. Therefore, cAMP-dependent kinases should be involved in the phosphorylation cascades they induce. The question arises: Does disulfide isomerization (cyclic reduction/oxidation with rearrangement) change the activity of plasma membrane receptors? Evidence for these changes has been found in the post-translational activation of the plasma membrane receptors for insulin, epidermal growth factor, and acetylcholine.[594] The cystine-rich regions of these three very different receptors are all located in the extracellular binding domain. Therefore, if the arrangement of the disulfide cross-links is essential for binding and activation of the receptors, they could be easily altered by reduction and rearrangements. Inserting an active receptor in the plasma membrane would require the proper post-translational alignment of the cross-links. The insulin receptor is a tetramer of two alpha subunits joining two beta subunits by disulfide (cystine) bonds.[594] The amino ends of all four subunits are on the extracellular side of the plasma membrane. The two beta subunits contain the transmembrane domains and the cytoplasmic tyrosine kinase domains with the autophosphorylation sites (tyrosines) near their carboxyl ends. The insulin receptor is a glycoprotein capable of binding Con A and wheat germ agglutinin (WGA). The insulin receptor is synthesized as a large, mannose-rich protein (190- to 210-kD M_r) that undergoes proteolytic processing and finishing of the N-linked carbohydrate chains to yield both the alpha and beta subunits. Three hours are required post-translationally for the receptor to appear at the cell membrane. Much of the processing is in the endoplasmic reticulum. The M_1 protein is the first product of translation (185 kD). It has a half-life of 15 min for conversion into M_2 (200 kD). This change involves intramolecular rearrangement of disulfide bonds.[594] The M_2 protein is converted by disulfide crosslinking into a dimer D, which is in turn converted to the complete receptor by proteolysis and polysaccharide maturation. The receptor, upon translation, lacks insulin-binding ability. The process required to obtain binding ability requires about 90 min and occurs in the endoplasmic reticulum. The half-life for obtaining this capability is 45 min. Additional processing time, which involves N-linked oligosaccharide modifications, dimer formation, and transport to the cell surface, adds to the 90 min to give a total of 3 hours. These latter processes occur in the Golgi apparatus. Post-translational oxidation and cross-linking of vicinyl cysteine residues occur nearly simultaneously with translation. These disulfide bonds are most likely not appropriate for ligand binding. In the endoplasmic reticulum, protein disulfide isomerase (glutathione:insulin transhydrogenase; EC 1.8.4.2) may perform this function or thioredoxin may do so. The former is localized to the endoplasmic reticulum.[594]

The epidermal growth factor (EGF) receptor is also a transmembrane protein, but it is a monomer of 170 kD.[594] It has one membrane-spanning domain separating the receptor into an extracellular N-terminus and cystine-rich domain and an intracellular carboxyl terminus domain with tyrosine kinase activity and tyrosine autophosphorylation sites. The extracellular domain contains the EGF-binding site. The EGF receptor is synthesized as a 160-kD precursor with high concentration of attached mannose. It is converted to a 170-kD receptor by maturation of most of its N-linked oligosaccharide chains. The EGF receptor acquires its EGF-binding capability within a half-life of 30 min. The remainder of the processing time leads to transportation to the cell surface in 1.5 hours. The EGF receptor has a half-life of 10–15 min to acquire autophosphorylating ability. Therefore, this activity is also acquired in the endoplasmic reticulum.

The acetylcholine receptor is quite different from the previous two receptors. It is composed of five subunits: two alpha, one beta, one gamma, and one delta.[594] These subunits compose a pentamer channel through the cell plasma membrane. The two alpha subunits are separated in the structure. Each alpha subunit has one high-affinity acetylcholine binding site (for a total of two per receptor). Each of these alpha subunits contains four cystine residues involved in the disulfide cross-links. These subunits show homology of 19 and 20% of four and three identical amino acid residues, respectively, and conservative substitutions in 54% of the other residues. They each have an amino-terminus extracellular domain and five membrane-spanning alpha helices. Four of these are hydrophobic, and one is amphipathic—contributing to the central ion channel. The subunits cannot be substituted for each other in the receptor. All the subunits are glycosylated at translation. The alpha and beta subunit associations occur in the endoplasmic reticulum. The oligosaccharides of the alpha and beta subunits are rich in mannose and can be cleaved by endoglycosidase H. Gamma and delta subunits contain complex oligosaccharide chains. Newly synthesized receptors cannot bind acetylcholine, but acquire the property during processing with a half-life of 40 min. Subunit assembly has a half-time of 80 min. Terminal glycoslylations and appearance of the receptor at the plasma membrane surface has a half-life of 2 hours.

Acquisition of ligand binding during post-translational processing has also been observed for the fibronectin receptor.[594] The initial glycosylations of the EGF, insulin and acetylcholine receptors may prevent aggregation and misfolding prior to proper disulfide cross-linking.

Evidence does exist that ties mature receptor activity to thiol/disulfide modification. The release of fluid into the intestinal lumen in suckling mice by treatment with heat-stable (ST) *E. coli* enterotoxin is mediated by cGMP.[596] ST activates a receptor that in turn upregulates guanylate cyclase. Cystamine and L-cystine (0.5 μmol per mouse) decreased intestinal secretion stimulated by 1.6 U of ST by 38 to 43%, respectively. At 1 mM, they decreased induced but not baseline guanylate cyclase (induced with 6 U/ml ST) by 33 to 73%. Cysteamine, cysteine, and acetylcysteine decreased secretion by 70 to 109% under the same conditions and decreased guanylate cyclase activity only 20–25% at 1 mM and 38–57% at 10 mM. Cystathione, which could not be reversibly oxidized and reduced, had no effect on either assay. None of these thiocompounds had an effect on 8-bromo-cGMP-mediated effects that bypassed the receptor and guanylate cyclase stimulation. ST does contain six cysteinyl residues and is inactivated by 2-ME and dithiothreitol. Also, the receptor binding itself may be modified by the thiols. Modification of the ST by the thiols used was unlikely since they must be reduced first to have action on the guanylate cyclase. Also, the thiols were more effective in decreasing secretion than the disulfides but less effective against guanylate cyclase, suggesting that a receptor effect was more likely with the exogenous thiols and that reduction of disulfides may have led to targeting to the guanylate cyclase (intracellularly).

Other correlates to this redox (thiol/disulfide) surface effect are found in the immune system. Benzene toxicity, marked by lymphocytopenia and hypoplasia in bone marrow, thymus, and spleen, correlates with the formation of polyhydroxy derivatives of benzene.[597] Rat splenic lymphocytes show enhancement of their *in vitro* phytohemagglutinin (PHA)-stimulated mitogenesis at low concentrations of hydroquinone (10^{-7} M), with a maximum effect at 14 hours, and inhibition of mitogenesis at >1.3 μM hydroquinone (maximum effect without cell killing at 10^{-5} M for 30 min).[598] Parabenzohydroquinone was a more potent inhibitor than 1,2,4-benzenetriol or catechol.[598] Dithiothreitol protected lymphocytes from these effects of hydroquinones. These effects are very similar to those observed with 3-amino-L-tyrosine, which is likely to form a quinonoid active form in cells.[218,562] *N*-Ethylmaleimide, a thiol-interacting agent that is taken up by cells, inhibited lymphocyte mitogenesis at the same

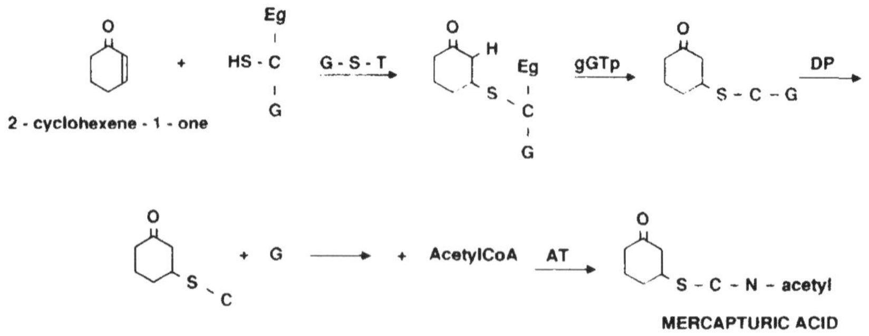

FIGURE 15. The formation and metabolism of the glutathione conjugate of 2-cyclohexene-1-one. Eg = gamma-glutamate residue; HS-C = cysteine; G = glycine; G-S-T = glutathione-*S*-transferase; gGTp = gamma-glutamyl transpeptidase; DP = dipeptidase; AT = acetyltransferase.

concentration as hydroquinone.[598] 5,5′-Dithio-bis(2-nitrobenzoic acid) (DTNB), which does not penetrate cells, had no effect on lymphocyte mitogenesis in these studies.

Evidence derived from studying human peripheral blood lymphocytes treated with the specific glutathione-conjugating agent 2-cyclohexene-1-one shows that thiol/disulfide receptor interaction plays a role in lymphocyte mitogenesis (Figure 15).[406] Concentrations of this reagent up to 5×10^{-5} M did not significantly affect the viability of lymphocytes or accessory cells like monocytes. A nearly 100% inhibition of tritiated thymidine incorporation was noted at the highest tolerated concentration of 2-cyclohexene-1-one with 0.75 µg/ml of PHA or 20 µg/ml of Con A per milliliter treatment in 72 hours. The ID_{50} for cyclohexeneone in the presence of PHA was 1.6×10^{-5} and in the presence of Con A it was 1×10^{-5} M. Cyclohexeneone at 2.5×10^{-5} M decreased GSH in treated lymphocytes to less than 20% of control cells and caused a 80 to 90% decrease in tritiated thymidine incorporation and blast transformation. The cyclohexenone was maximally effective only if provided to the cells within 4 hours of culture. After this time there was no effect. Other studies showed that T cells treated with Con A for 5 hours had more mixed glutathione-protein disulfides than control resting cells.[599] The mixed disulfide reaction has been suggested as part of the lectin/receptor transduction process in lymphocytes because cyclohexeneone-treated cells do not upregulate cAMP levels when treated with PHA or Con A as do control cells.[599] However, if these cyclohexeneone-treated cells are exposed to isoproterenol or prostaglandin E, they do produce cAMP. Therefore, the adenylate cyclase is not directly affected by the thiol-conjugating agent and the receptor is the more likely target.

Anti-lipid-peroxidation agents have also been shown in mouse spleen cells stimulated specifically with sheep red blood cells (plaque-forming lysis test) or stimulated nonspecifically with LPS (proliferation) to enhance immune responses.[583] Superoxide dismutase, butylated hydroxyanisole/butylated hydroxytoluene/*n*-propyl gallate, lucigenin, and alpha-tocopherol all enhanced both splenic cell responses described above.[583] These antioxidants were additive in effect with suboptimal concentrations of 2-ME. Stimulation of peroxidation in mouse splenic cells *in vitro* with particles of silica, talc, Bentonite, or *Corynebacterium parvum* cells inhibited LPS- and Con A-stimulated proliferation of these cells and sheep-red-blood-cell-stimulated antibody production after at least 48 hours of exposure.[600] Both alpha-tocopherol and 2-ME prevented this inhibition.[600]

The surface effects of oxidation, reduction and thiol-binding agents on lymphocytes and macrophages have been related to microtubular and microfilament assembly and disassembly (cytoskeletal effects) and receptor movement and membrane transport mechanisms.[582,598] Soluble immune response suppressor is activated by macrophages or hydrogen peroxide to the active oxidized form.[601-604] This form inhibits cytoskeletal-dependent functions in lymphocytes and enhances capping even in the presence of saturating concentrations of lectins.[605] Levamisole, thiabendzole, and sodium aurothiomalate interfere, perhaps by sulfide reductase processes, with nonspecific suppressor T cell activity.[582] Levamisole can protect against soluble immune response suppressor or interferon-mediated suppression of plaque-forming spleen cells and enhances the uptake of cystine in lymphocytes.[582,602]

Finally, in human peripheral blood T lymphocyte subpopulations (CD4$^+$ and CD8$^+$), GSH or protein thiol levels have been found to play an important role in the earliest events of immunostimulation.[406] Historically, to the study under discussion, GSH-depleted lymphocytes were shown to be still capable of IL-2 production and IL-2 receptor expression following lectin treatment (late events in G_1). However, early in human T cell activation (between G_0 and G_1) stimulation of growth by anti-CD3 monoclonal antibody did depend on GSH levels. Signal transduction has also been shown to be sensitive to redox and thiol states in endothelial cells, granulosa cells, and fibroblasts. In a 1993 study, 1-chloro-2,4-dinitrobenzene (CDNB), a glutathione-depleting agent by the action of glutathione transferase, yielded 51, 41, 3, and 0% T cells exiting G_1 when treated with anti-CD3 antibody and, respectively, 0, 5, 15, and 45 µg/ml of CDNB for 15 min at 37°C.[406] Treatment with CDNB also caused a dose-dependent decline in free intracellular calcium ions for both CD4$^+$ and CD8$^+$ cells. CD4$^+$ showed a much higher sensitivity for intracellular calcium ion decline. The effect on calcium mobilization could not be reversed by dithiothreitol treatment. However, this result is not surprising because the *S*-(2,4-dinitrophenyl)-glutathione (DNPG) conjugate, resulting from the glutathione-*S*-transferase activity on CDNB, inhibits oxidized glutathione (GSSG) transport and reduction in liver and red blood cells.[606] Nitroaromatic compounds are known to be strong inhibitors of glutathione reductase.[406,599] For short exposures (less than 30 min), DNPG and CDNB were reversible inhibitors of glutathione reductase at K_i's of 22 and 30 µM, respectively.[406,599] After exposures of more than 30 min, CDNB irreversibly (uncompetitively) inhibited glutathione reductase. The rate of this uncompetitive inhibition is related to the GSSG levels; therefore, the CDNB probably covalently interacts with an intermediate state of the enzyme. Nitrofurantoin is a competitive inhibitor (K_i = 20 μM) of GSSG reductase and prevents inhibition by DNPG.[606]

CDNB treatment (5 μg/ml) of CD4$^+$ cells caused a decrease in intracellular free calcium ions in both early and late phases of stimulation, while in CD8$^+$ cells it caused only a late-phase decrease.[406] Larger doses (15 and 45 μg) affected both cell types in both phases equally. Only simultaneous treatment with dithiothreitol could prevent the effect of CDNB.

Another event early in T cell receptor (CD3) activation is tyrosine phosphorylation.[403,607] Specifically, tyrosine kinase p56 lck and fyn are activated. This event precedes phospholipase C-gamma-1 (PLC-gamma-1) activation (a major driver of calcium mobilization).[608] The profile of phosphorylation was shifted in CDNB-treated T cells.[406] Protein substrate-phosphorylation of 135-, 110-, and 35-kD proteins was inhibited while 70- and 80-kD substrates showed increased phosphorylation. PLC-gamma-1 and a coprecipitated 35-kD protein decreased in phosphotyrosine content with CDNB treatment. At intermediate CDNB concentrations, increased phosphorylation of tyrosine was noted in 60-, 70-, and 110-kD proteins of lymphocytes. However, at the high concentrations, they showed a decrease in phosphorylation.

Previously, low GSH levels have been seen to inhibit mitogen-stimulated RNA synthesis in lymphocytes.[406,572,582,599] Other studies have shown that treatment of cells with BSO, a GSH-synthesis inhibitor, blocks G_1/S transition in the cell cycle.[572,599] CD4 and CD3 receptor interaction is essential for initiating protein phosphorylation within seconds of stimulation.[403] The transduction through CD4 is mediated by tyrosine kinase p59 lck.[607] CD3 transduction is mediated through tyrosine kinase p59 fyn.[607] CD4 regulates CD3-mediated tyrosine phosphorylation of PLC-gamma-1 and, therefore, calcium mobilization. Both CD4 and CD8 alpha receptors contain extracellular N-terminal cysteine groups (like the nonimmune cell receptors discussed earlier), which are critical to p56^{lck} tyrosine kinase activity. Again, the dependence of these activities on intracellular GSH and the location of the kinase-controlling-receptor cysteines extracellularly suggest that the extracellular disulfide reductase must be intramembranous or a transmembrane protein.

VII. STRESS GROWTH, INHIBITORY AND INFLAMMATORY FACTORS

A. NATURAL GLUTATHIONE CONJUGATES

At the beginning of this chapter, observations of the feeder cell phenomenon suggested that the extracellular disulfide reductase effects of macrophages and other cells are mediated strictly through thiols. However, evidence that follows will suggest that conjugates of thiols and costimulators also play a very important role in the functions of the extracellular disulfide reductase activity.

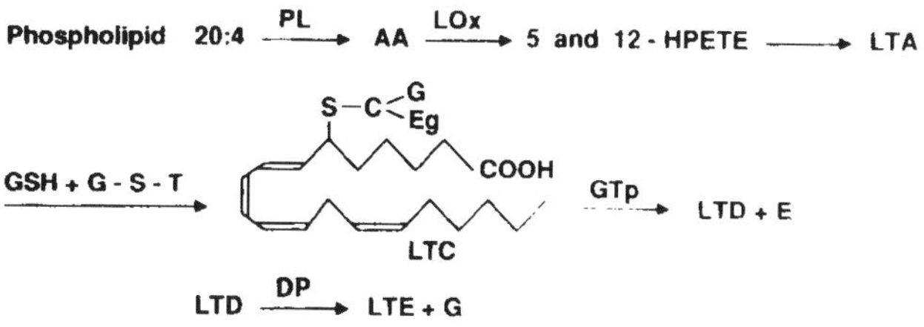

FIGURE 16. Macrophage leukotriene synthesis. PL = phospholipase; AA = arachidonic acid; LOx = lipoxygenase; 5 and 12-HPETE = 5- and 12-hydroperoxyeicosatetranoic acids; LTA = leukotriene A; GSH = reduced glutathione; G-S-T = glutathione-S-transferase; C–S– = cysteinyl residue; G = glycine; Eg = gamma-glutamyl residue; LTC = leukotriene C; GTp = glutamyl transpeptidase; LTD = leukotriene D; E = glutamate; DP = dipeptidase; LTE = leukotriene E.

The principal natural conjugate of glutathione is leukotriene C (LTC).[609,610] Originally, this material was described as a component of the slow-reacting substance of anaphylaxis (SRS-A), along with leukotrienes D (LTD) and E (LTE). SRS-A produces the sustained bronchioconstriction in humans associated with antihistamine-resistant symptoms of asthma. LTD is 3–20 times more vasoactive than LTC, and LTE exhibits very low activity.

Figure 16 demonstrates the synthetic pathway for production of SRS-A in macrophages.[610] Leukotrienes C, D, and E are synthetically derived from each other in their alphabetical order. These metabolites peak around 60 min following stimulation and may be maintained in steady state at this level for in excess of 2 hours.[609] Macrophages bind IgG and IgE complexes by Fc receptors and synthesize LTC in response to this stimulation, albeit at lower concentrations than are obtained when they are stimulated with opsonized zymosan. Pretreatment of mice with *C. parvum* yielded peritoneal macrophages that did not release LTC upon opsonized zymosan stimulation. LTC release by mouse pulmonary interstitial macrophages is greater than that of mouse pulmonary alveolar macrophages.[610] Even though pulmonary interstitial macrophages release less 20:4 phospholipid than peritoneal macrophages upon stimulation, the pulmonary macrophages form predominantly lipoxygenase products while the peritoneal macrophages form predominantly cyclooxygenase products (prostaglandins, thromboxanes, and prostacyclin).

B. XENOBIOTIC GLUTATHIONE CONJUGATES

Xenobiotic glutathione conjugates have already been alluded to in the cases of 2-cyclohexene-1-one and 1-chloro-2,4-dinitrobenzene. The glutathione-*S*-transferase activity that produces these conjugates can be envisaged as a detoxifying process, increasing solubility and excretion and blocking interactions with critical functional thiols in proteins, or as a toxifying process, increasing availability and active site interaction for enzymes requiring exogenous disulfides or thiols.[560,611,612] As expected, both processes are demonstrable. In fact, the oxidized glutathione (disulfide) can be viewed as a self-conjugate that is toxic in itself.[560,566] This fact is why the export of glutathione disulfide appears to be universal in mammalian cells.[566]

To examine these processes more closely, two organs in which xenobiotic/thiol metabolism has been intensely studied must be considered. These organs are the liver and kidney.[566] Liver is the organ principally responsible for extracellular, or plasma, levels of reduced glutathione (in the micromolar range) compared to tenths of millimolar to millimolar range within cells.[566] Liver cells actively export reduced glutathione, but like many other tissue cells only poorly take it in. The hepatocytes are deficient in the membrane-bound gamma-glutamyltransferase (EC 2.3.2.2; (5-glutamyl)-peptide:amino acid 5-glutamyltransferase), which is responsible for the removal of the glutamate residue and transfer of other amino acids to it with release of the cysteinyl-glycine dipeptide. The cysteinyl-glycine is then broken down into cysteine and glycine by cysteinyl-glycine dipeptidase (EC 3.4.13.6; L-cysteinyl-glycine hydrolase). This last enzyme occurs in renal tissue but not hepatocytes. Also, hepatocytes are poor transporters of cystine and other disulfides (like lymphocytes). Consequently, hepatocytes use plasma methionine as a substrate for synthesizing cysteine and eventually GSH. Cysteine can be taken up by hepatocytes and used to make GSH, but extracellular autoxidation limits it availability. Methionine does not support GSH synthesis in renal cells even though it is readily taken up by them. This inability is probably based on the low cystathionase (EC 4.4.1.1; L-cystathione cysteine-lyase) (deaminating) activity in renal cells.

Renal cells on the other hand have several strategies for taking up GSH and its components.[566] The gamma-glutamyltransferase is one that has already been mentioned. Cystine is also readily taken up but its reduction to cysteine for utilization in glutathione synthesis is, in itself, dependent on cytosolic thiol transferase and glutathione reductase (EC 1.6.4.2; NAD(P)H:oxidized-glutathione oxidoreductase). GSSG, in itself, is readily taken up by renal cells. Other disulfides are also taken up by renal cells, as illustrated by the renal protectant mercaptoethane sulfonate. It is not taken up or reduced by hepatocytes. However, renal cells take it up from the plasma as a disulfide and excrete it into the urine as a thiol. This agent is used to protect the urinary tract from toxicity resulting from treatment with oxazophosphorine cytostatics.

The gamma-glutamyltransferase and dipeptidase are located in the proximal tubular epithelial brush border.[566] However, another type of glutathione metabolism exists in renal cells other than that performed by these enzymes. It involves the uptake of GSSG and its subsequent reduction intracellularly or degradation to constitutive amino acids by hydrolase. The conversion of extracellular GSH to GSSG prior to uptake by a translocase is accomplished by a renal thiol oxidase (copper-dependent) in the plasma membrane of tubular epithelial cells. Hydrogen peroxide is formed in the oxidation of GSH to GSSG. This oxidase acts on extracellular thiols such as GSH, cysteine,

N-acetylcysteine, and dithiothreitol. The enzyme is located in the basolateral part of the tubular epithelial plasma membrane. Therefore, it acts primarily on plasma thiols. However, the gamma-glutamyltranspeptidase may still be the principal enzyme in renal GSH metabolism because much of the plasma glutathione is removed from the plasma by transtubular transport.

Because the liver may be subject to considerable oxidative stress from superoxide leakage from cytochrome P450 metabolism of drugs and toxins to nitric oxide production and subsequent superoxide production from uncoupled mitochondrial respiration, GSSG may accumulate faster in liver cells than it can be reduced. The depletion of NADPH by competing redox chains (such as glutathione reductase/peroxidase, fatty acid synthesis, and cytochrome P450 activity) leads to excretion of GSSG by hepatocytes.[566] The hepatic GSSG translocase is located primarily in the plasma membranes of hepatocytes that form the bile canaliculi. The net removal of oxidants by superoxide dismutase, catalase, and glutathione peroxidase leads to altered calcium metabolism in hepatocytes. The loss of calcium ions from the nonmitochondrial compartments (mostly endoplasmic reticulum) is oxidant dependent and thiol preventable.[366] The loss of calcium ions from the mitochondria is based on NAD(P)H depletion. Examination of ATP-depleted liver microsomes showed that calcium-dependent-sequestration inhibition could be removed by addition of dithiothreitol or GSH. Direct oxidant damage, conjugation of GSH, or release of mitochondrial uncouplers from GSH conjugates could lead to calcium release and subsequent cell death.

Detoxification and intoxication targeting of a variety of xenobiotics and pharmaceutical agents to the liver and renal tissues are based on the aforementioned metabolisms. The initial event in most tissues is carried out by glutathione transferase (EC 2.5.1.18; RX:glutathione R transferase).[560,566] This enzyme is inducible. Hepatic and renal tissue drug metabolism differences can be illustrated by the metabolism of the analgesic paracetamol.[566] Although the major liver metabolites are sulfate and glucuronic acid conjugates, the electrophilic metabolite formed by cytochrome P450 monooxygenase system (probably an imine quinone) reacts with glutathione by transferase activity. The inactive glutathione-*S*-paracetamol is converted to mercapturate by gamma-glutamyltransferase, dipeptidase, and *N*-acetyltransferase activities before being excreted in the urine. If GSH is consumed by oxidative stress or some other conjugate during paracetamol metabolism, the alkylating electrophile of paracetamol will kill hepatocytes.

In liver cells, the paracetamol that is glutathione conjugated is not converted to the mercapturate. In renal cells, the mercapturate is formed. The hepatocytes readily release the glutathione conjugate to be further metabolized by renal cells with the appropriate enzyme.

Halohydrocarbons can be activated to toxic species by the interaction of glutathione and glutathione-*S*-transferase.[566,611-613] For instance, glutathione and vicinal-dihaloethanes are converted into sulfur mustards that alkylate

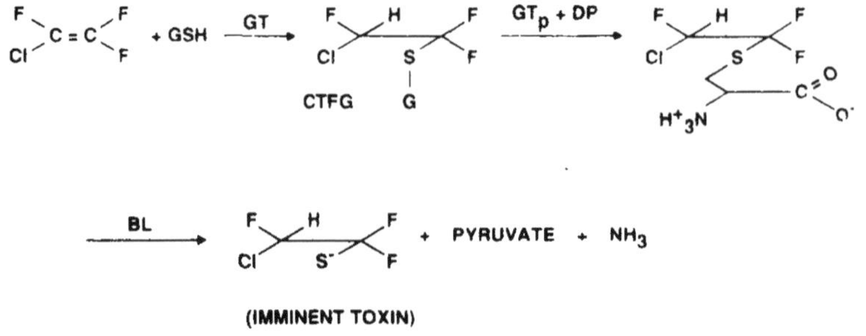

FIGURE 17. *Glutathione conjugation and metabolism of chloro-1,1,2-trifluoroethylene. GSH =* reduced glutathione; GT = glutathione-*S*-transferase; CTFG = *S*-(chloro-1,1,2-trifluoroethyl)-L-cysteine; GT_p = glutamyl transpeptidase; DP = dipeptidase; BL = beta-lyase.

DNA. Haloalkenes are converted to the corresponding *S*-glutathione conjugates, which are then metabolized to cysteinyl derivatives by gamma-glutamyltransferase and dipeptidase. Subsequently, the renal cysteine conjugate beta-lyase converts the cysteinyl derivative to pyruvate, ammonia, and a reactive thiol that is the imminent toxin. The result is nephrotoxicity.

An example of such a process is seen in the metabolism of chlorotrifluoroethylene, which is highly nephrotoxic. It is converted to the glutathione-*S*-conjugate *S*-(2-chloro-1,1,2-trifluoroethyl)-glutathione (CTFG).[566,611] The reaction is more actively catalyzed by liver microsomal glutathione-*S*-transferase than liver cytosolic or renal microsomal or cytosolic enzymes. Other haloalkenes processed in this way by the liver include chloro-1,3-butadiene and tetrafluoroethylene into *S*-(pentachlorobutadienyl) glutathione and *S*-(1,1,2,2-tetrafluoroethyl) glutathione, respectively.[612]

CTFG is further metabolized by renal tissue gamma-glutamyltransferase into the *S*-(chloro-1,1,2-trifluoroethyl)-L-cysteine (CTFC) (see Figure 17).[566] The gamma-glutamylytransferase inhibitor acivicin can protect renal cells from the toxicity of this agent. Other toxins that are metabolized in renal tissue in this same way are *S*-(1,2-dichlorovinyl) glutathione and *S*-(pentachlorobutadienyl) glutathione. CTFC is further activated in renal tissue by pyridoxal phosphate-dependent beta-lyase. The selective inhibitor of pyridoxal phosphate-dependent enzymes aminooxyacetic acid can protect against the toxicity of this agent at this level. CTFC and *S*-(1,2-dichloro-vinyl)-L-cysteine (DCVC) are powerful renal mitochondrial toxins, affecting the mitochondrial redox chain and Krebs cycle enzymes.

The processes of detoxification vs. toxin activation by glutathione conjugation are further complicated by the process of toxin reactivation.[612] This latter process involves glutathione interaction with glucuronide derivatives of toxins. Glucuronic acid conjugation is the principal detoxification pathway in liver for nonsteroidal anti-inflammatory agents, hypolipemics, analgesics, diuretics, acetylcholinesterase inhibitors, aldose reductase inhibitors, pesticides, herbicides,

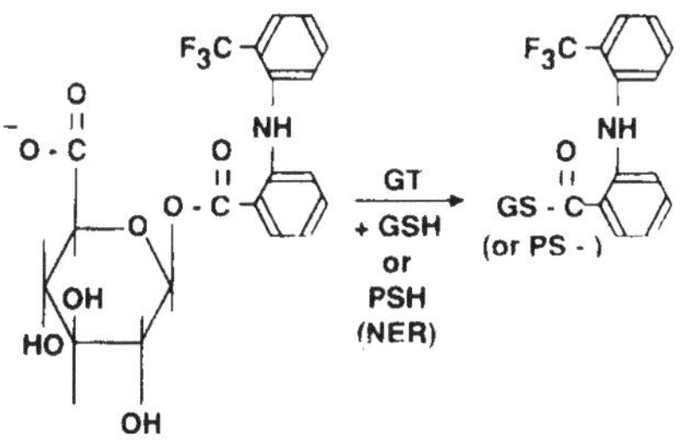

FLUFENAMIC ACID
GLUCURONIDE

FIGURE 18. Retoxification of flufenamic acid glucuronide. GT = glutathione-*S*-transferase; GSH = reduced glutathione; PSH = protein thiol; NER = nonenzymatic reaction, transfer to protein thiol; PS– = protein thiol conjugate.

and plasticizers. The problem of retoxification arises with microsomal glutathione transferase. Figure 18 shows the displacement of glucuronic acid from flufenamic acid glucuronide by glutathione mediated by the transferase or nonenzymatic displacement by protein sulfhydryl groups (which is a much slower, chronic reaction).[612] Serum albumin is a marker recipient of this transfer.[612] Other substances and drugs that can participate in such a transfer are indomethacin, clofibric acid, bilirubin, benoxaprofen, and chlorambucil. Such transacylations may involve nucleophiles other than thiols, such as hydroxyl and amino groups of proteins, nucleotides, and carbohydrates. Because of the slow reaction rates or reversibility of these processes, toxicity may be low or nonexistent. However, in the case of the hypolipidemic drug clofibrate, the results were lethal. An international double-blinded study of its use showed that more chronically treated patients died from side effects of the drug, including liver and renal necrosis and carcinogenicity, than from cardiovascular disease if the treatment had not been given.[612]

The depletion of glutathione by conjugation or the alteration of protein by xenobiotic conjugation of critical sulfhydryls may be two mechanisms of chronic toxicity by the transacylations.[613] Another, more obscure effect of the conversion of GSH to GSSG or GSX is the disruption of microtubular assembly from tubulin, which requires free sulfhydryl groups and is inhibited by rises in GSSG.[598] Vinca alkaloids, which disrupt this assembly, also increase the accumulation of GSSG in cells by an unknown mechanism.[614] In studies using H-35 rat hepatoma cells, vinblastine sulfate caused a marked increase in intracellular levels of GSSG (from 0–24 pmol/mg protein at 0 time to 1736 pmol/mg protein at 4 hours). Reduced glutathione levels in these cells were held at steady state. Vincristine and vindesine produced the same results, but not adriamycin and methotrexate. Protein synthesis was inhibited as GSSG levels rose. Mixed disulfides of glutathione are toxic. The depletion of GSH does not just make cells vulnerable to reactive oxygen intermediates but leads to the toxic effect of these mixed disulfides that can be chronic.[614]

Cataract formation provides an example of both actions. Treatment of adult mice with the inhibitor of glutathione biosynthesis, L-buthionine sulfoximine (L-BSO), results in nontoxic effects.[615] In suckling mice 9 to 12 days old, no lethal or chronic injurious effects were observed except for dense cataract development.[615] In mice 14 to 17 days old, treatment with L-BSO resulted in death, hind-leg paralysis, and/or impaired spermatogenesis but no cataracts.[615] In senile cataracts in humans, there is an age-dependent decrease in GSH and an increase in glutathione/protein mixed disulfides. This decrease in GSH occurs as a consequence of aging, but is even more severe with cataract formation.

In advanced senile cataracts of humans, GSH bound to proteins increases.[616] In advanced clinical cataracts in various breeds of dogs, GSH bound to proteins was increased.[616] This was also true for lenses with congenital cataracts from adult miniature schnauzers.[616] In congenital cataracts of pups, soluble protein-bound GSH was lower than in lenses of normal pups. These dynamic changes in GSH and bound GSH indicate various stages in the cataractogenesis process.

Another major thiol source in cells is metallothioneins, which play an important role in complexing with group IIB metals, in zinc and copper ion homeostasis, in heavy and transition metal detoxification, and in detoxifying oxygen radicals.[617] They also provide cell resistance to antineoplastic drugs such as cis-diaminedichloroplatinum (II), but not 5-fluorouracil or vincristine.[617] They also provide resistance to the effects of the alkylating agents chlorambucil and mepholon. The metallothioneins are induced (transcriptionally) by metals, epinephrine, glucocorticoids, hyperthermia, cytokines, cyclic nucleotides and phorbol esters.

C. VIABILITY FACTOR

As noted earlier, the thiols are intimately linked to protein synthesis, detoxification, metabolic activation, transport, membrane receptor activation, and second messenger effects. What do these observations tell us about the nature of the apparent extracellular disulfide reductase and its relationship to cytochrome b systems? Thiols are responsible for maintaining the functionality of NO synthase, NADPH oxidase, and their long-range effects. For instance, NO is transported as nitrosothiol, and oxidations of critical thiols by reactive oxygen intermediates lead to metabolic modification and synthetic changes in cells (if not death). It is apparent that many polypeptide hormones, receptors, and transport functions require thiol modifications or maintenance for their operation. In fact, the polypeptide hormones themselves are excellent protein disulfide reductases, or isomerases. But where are the reducing equivalents for the extracellular disulfide reductase or isomerase activities coming from? Although certain cell types such as hepatocytes can reduce and export glutathione, glutathione in itself, without a transhydrogenase, is not a good reducer of protein disulfides. Cysteine, which is readily autoxidized extracellularly, lacks the reducing power to directly reduce protein disulfides efficiently.

Examining the interaction of autocrines and cytokines with extracellular thiol production provides a clue to how reducing equivalents may be provided transmembranously to the polypeptide (hormone) reductase/isomerase bound to its respective receptor.

A critical question is what triggers the overshoot production of extracellular, or exported, thiols. Although extracellular thiol production can be tied to amino acid transport systems, as noted earlier, this mechanism is more likely to produce balanced exchange than overshooting production. Regardless of whether a cytochrome b is involved with the actual transmembrane transhydrogenase reaction or not, the two cytochrome-b-based oxidative systems—NO synthase and NADPH oxidase—are likely to provide the impetus for an increased thiol response.

Observations of mononuclear leukocytes in culture support the hypothesis that increased endogenous oxidations drive the disulfide reductase activity. Treatment with LPS (0.02 to 2 μg/ml), opsonized zymosan, and glucose oxidase-horseradish peroxidase colloidal particles increased the reactive oxygen metabolites of RAW 264.7 mononcyte/macrophage cells *in vitro*.[574] These treatments caused a 25% (for oxidase-peroxidase) to 100% (for LPS and zymosan) increase in thiol release by RAW cells in 24 hours of incubation.[574] In contrast, P388D$_1$ mouse macrophage cells that did not respond oxidatively to LPS treatment did not show increased release of thiols. Furthermore, 3-amino-L-tyrosine, which blocks the oxidative burst of macrophages, decreased thiol release by RAW 264.7 cells by 36–57%.[574]

TNF, usually associated with stimulation of destructive oxidative and mitochondrial metabolism in macrophages, neutrophils, hepatocytes, and target tumor cells, also provides growth stimulation for oxidatively sensitive T cells and thymocytes. Recombinant human TNF alpha and beta (1000 U/ml) stimulated a sixfold increase in DNA synthesis in the mixed peripheral human lymphocyte reaction.[618] Furthermore, TNF-alpha was found in the supernatants of cells undergoing this reaction within 1 hour of its initiation.[618] This TNF-alpha peaked in concentration at 4 hours after stimulation. Antibodies, specific to TNF-alpha, inhibited the mixed lymphocyte reaction. These cells were oxidatively very sensitive and required 5×10^{-5} M 2-ME in their culture medium. The TNF needed to be added with the first 3 days of culture to gain the enhancement. Therefore, it was involved in an early event in the reaction. IL-2 receptor expression increased with the TNF stimulation.[564] This receptor has shown the requirement for a thiol-generating costimulator (thioredoxin). In CTLL-2 cells in culture, IL-2 stimulation leads to accelerated death of the cells unless 2-ME is present.[520] IL-1 induction by TNF in the mixed lymphocyte cultures may also play a role in the proliferation response.[618]

The linkage between oxidative stimulation, thiol production, and proliferation suggests that if the thiol-generating system is incapacitated, then the same stimulus that caused proliferation will also accelerate death of the cells when the thiols are absent. Therefore, extracellular thiols would play a critical role in the growth or death switching. Again, examining the stress responses of RAW cells provided evidence for such a switching mechanism.

When stimulated with high-cell-density stress (30 min at 5.7×10^7 cells per milliliter) or LPS alone or with normothermic microwave radiation (100 m W/g, 2450 MHz at 37°C), RAW cells first show growth inhibition, then increased extracellular thiol production, followed by decrease in thiol production and resumed growth.[218] Simple transfer manipulation of RAW cells leads to a 50% loss of viability (measured by trypan blue).[218] However, transfer of supernatants (conditioned medium) from stressed cells (treated as above) increased viability of transferred RAW cells to over 90%.[218] The viability response is dose dependent on the amount of fresh culture medium replaced by stressed-cell-conditioned medium. Preliminary partial purification by ion exchange chromatography and dialysis shows that the transfer factor is a protein.[355] The autocrine present in the RAW conditioned medium remains unidentified. This high-cell-density stress or LPS treatment not only increased thiol production and release of an autocrine, but also led to increased NO synthase activity.[521] During the 30-min stress period, no increase in messenger RNA transcription was noted for c-fos, c-jun, mouse sarcoma virus, Hsp70, or TNF.[218] Therefore, the hysteresis that affected growth and metabolism was mediated through an epigenetic event that persisted after removal of the stressor. Brief exposure to 3AT simultaneously with the LPS increased both cell growth and thiol production over treatment with LPS or 3AT alone.[218]

These cells, which were originally transformed with Abelson leukemia virus, are latently infected and supposedly nonpermissive for the expression of the retrovirus.[619] However, treatment with as little as 0.1 ng/ml LPS or high-cell-density stress led to expression of the virus and release of intact Abelson leukemia virions (3–4 days postexposure).[351]

A polypeptide cytokine originating from nonimmune tissue that similarly controls macrophage function through lymphocyte action has been identified. A 1988 study by Bernton et al. showed that hypoprolactinemic mice have suppressed macrophage activation and T lymphocyte function.[620] Furthermore, cyclosporine, an inhibitor of T cell function, inhibits binding of prolactin to lymphocytes.[620] Agents that normally induce oxidatively active macrophages in mice, *Mycobacterium bovis* (BCG), *Listeria monocytogenes,* and *Proprionibacterium acnes*, failed to induce tumoricidal macrophages in mice treated with bromocryptine. The proliferative response of splenic T cells to PHA was also suppressed by chronic bromocryptine treatment of mice. Bromocryptine is a dopamine type-2 agonist that inhibits the release of prolactin from the pituitary gland. Treatment of the mice with ovine prolactin reversed this effect. Gamma-interferon production by T cells was most impaired by the prolactin deficiency (treatment with graded doses of bromocryptine for 4 days). Splenic lymphocyte proliferation stimulated by ConA or LPS *in vitro* was also inhibited in cells from these mice. The treatment of the mice with prolactin (20 µg/day) on days 3 and 4, partially reversed the inhibition of proliferation. The lack of prolactin did not selectively affect subpopulations of splenic lymphocytes (B and T cells). Bromocriptine had no direct effect on macrophage function *in vitro*. Gamma-interferon and macrophage activating factors (undefined lymphocyte-conditioned medium) reversed the inhibition in macrophages induced by bromocriptine treatment of the mice. Splenocytes from bromocriptine-treated mice produced one fifth the gamma-interferon of untreated mice. Prolactin treatment restored the gamma-interferon production of these lymphocytes. Also, more animals died that were infected with Listeria and treated with bromocriptine than those that did not receive the drug or that received it but were treated with exogenous prolactin. These responses confirmed that the immunosuppression was due to prolactin deficiency. As mentioned earlier, prolactin and its receptors have been associated with a redox transduction process.

The most direct biochemical evidence that a redox process is involved in transduction and gene expression is seen with regulation of c-fos and c-jun binding to DNA.[592] The monomers dimerize through leucine zipper domains. The DNA-binding property originates from basic amino acid residues contributed by both proteins. Optimal binding of the heterodimer to DNA depends on a single hyperreactive cysteine in each of the monomers. If these cysteines are oxidized, binding is decreased. The oxidation may be reversibly resulting in a sulfenic or sulfinic acid. The AP-1 site on the DNA is also in juxtaposition with the cAMP responsive elements. A nuclear factor from the liver that facilitates the heterodimer formation and DNA binding has been discovered and requires a thiol for activity. Substitution of the hyperactive cysteine, designated C_1, with serine increased the DNA binding capability of the heterodimer. However, replacing C_1 with methionine decreased or eliminated DNA binding. This nuclear stimulating factor was temperature dependent, showing decreased activity at 25°C and none at 4°C. Thioredoxin stimulated the nuclear extract stimulatory factor. Thioredoxin showed the greatest activity when its reductase and NADPH were present. The concentration of nuclear stimulatory factor required for the same enhancement decreased from 2 to 0.1–0.05 µg, when the thioredoxin and its reductase were present. Thioredoxin itself did not directly affect binding of the c-fos-c-jun heterodimers to DNA. In v-jun (the oncogenic form), the redox regulation is not operable because the C_1 cysteine is replaced with serine, which as noted earlier increases DNA binding without the reductive process. The C_1 cysteine is located in a lys-cys-arg tripeptide, making it very reactive like the reductase/isomerase active sites of thioredoxin, protein disulfide isomerase, lutropin, and follitropin. The E_2 transcription factor of papilloma virus has a similar hyperreactive thiol. Enhancing transcription of viral genes with thiol-containing transcription factors and conditioned medium containing high concentrations of thiols/disulfides is the subject of the next section.

D. RETROVIRUS (HIV) ACTIVATING FACTOR

The discovery of altruistic factor(s), released from stressed RAW 264.7 cells, that rescue(s) unstressed cells from subsequent injury when stressed in conjunction with expression of latent Abelson leukemia virus suggested that transferable retrovirus activators exist.[218] The presence of enhanced extracellular thiol production in conjunction with this viral expression suggested a redox-linked process. To lend biochemical credance to this assumption, the tat protein, or transactivating protein, which upregulates human immunodeficiency virus (HIV) expression, has been found to be a protein of 86 amino acids containing 2 lysines and 6 arginines within 9 residues for nucleic acid binding and containing 7 cysteines within 16 residues.[593] The cysteines are always conserved within different isolates of HIV.[593] Both the basic amino acid and cysteine-rich regions are required for tat activity. Tat protein readily autoxidizes to mutimeric forms. The tat monomers form a metal-ion-cysteine-cross-linked dimer that is stable at pH 4.0.[593] Oxidized tat does not bind metal. Cadmium divalent cations are bound much more tightly than zinc divalent cations. Optical absorbance measurements have indicated that the binding affinities of the metals are in the order of cadmium > zinc > cobalt. Mercury and copper divalent cations and copper monovalent cations formed dimers of tat, but ferrous and ferric and divalent cobalt did not in electrophoresis assays. The tat dimer is composed of 14 cysteines that bind 4 metal ions.[593] Each metal ion is proposed to be surrounded by 4 cysteines with 2 metal ions and 8 cysteine thiolates acting as bridging ions between the monomers. Whether these cysteine and metal ions participate in a redox cycle or not is unknown. They may, like the iron ions and cysteines in repair endonuclease III, be involved in DNA binding. Nonetheless, oxidation would yield a nonfunctional tat protein. The human T cell leukemia virus (HTLV-1) of adult leukemia has transactivating factors similar to that of HIV (tat and rev for HIV and tax and rex for HTLV-1).[360,621,622] In human cells, there is a cellular trans-activation-responsive (TAR) element RNA-binding protein (TRBP) that works in conjunction with tat to upregulate latent HIV.[623,624] The cDNA for TRBP has been isolated from HeLa cells.[624] Presumably, such cellular transactivating proteins exist for HTLV-1 and other retroviruses. The basic amino acid residues of tat bind to the TAR sequence RNA and the long terminal repeat sequence (LTR) of the viral DNA incorporated in the host genome to initiate viral expression (transcription and translation).[624]

The conditioned medium of RAW 264.7 cells contains a transferable factor that upregulates HIV-1 without the benefit of tat.[625] Figure 19 shows the results of an LTR-CAT (chloramphenicol transferase bacterial gene attached to the LTR of HIV-1) expression assay in HeLa cells, demonstrating the upregulation of chloramphenicol transferase activity attached to the HIV-LTR promoter.[352,626] This factor, which is separable from the viability factor by anion exchange chromatography, is produced and released into media by RAW 264.7 cells along the same time course as the viability factor(s).[355] If this is the same factor

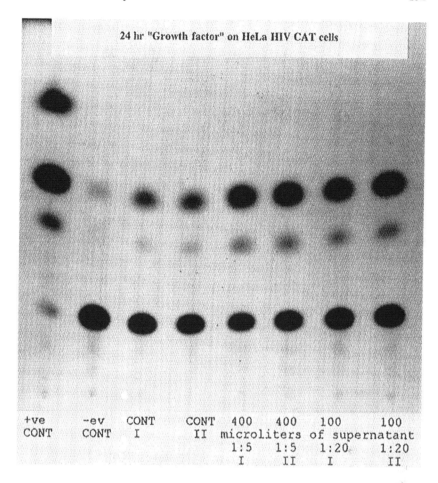

FIGURE 19. Upregulation of HIV-LTR-CAT in recombinant HeLa cells by 24-hour treatment with 1:5 (I and II) and 1:20 (I and II, duplicates)-dilutions of RAW 264.7 mouse monocyte/macrophage-derived cytokine. tve = positive control of chloramphenicol acetyltransferase changing solubility of ^{14}C-labeled chloramphenicol on thin layer chromatography plate (autoradiograph); -ve = negative control; controls I and II = spent media without RAW supernatants or extracts.

that upregulates the Abelson leukemia virus in RAW cells (3–4 days later), then the thiol production must follow the stress and the autocrine interaction but precede the viral expression. An initial oxidative event such as superoxide production or NO production would inactivate sensitive thiols contained in tat or a substitute cellular transactivating factor. Since no new mRNA was evident in key markers during the 30-min stress period and the factors (viability and retrovirus activating) were extractable from the cytoplasm of resting RAW cells, then the release and/or activation of these factors must follow a metabolic event and their activation of target cells must follow later events (24–72 hours later).[218,351] In clinical situations, the latency of HIV may, in fact, not be much of a cellular event as first thought. It may be based more on tissue sequestration.[627,628]

The germinal centers of lymph nodes of asymptomatically infected individuals have shown large numbers of CD4+ lymphocytes and macrophages latently infected with HIV.[629] A heavy association of HIV virions with the antigen-presenting follicular dendritic cells without intracellular HIV RNA or DNA has been found in these patients. These infected cells are sufficient to explain the ultimate lymphocyte depletion, resulting in symptomatic acquired immunodeficiency syndrome (AIDS). The dendritic cells of these nodes had many complete HIV virions in the interdendritic extracellular spaces of the cells. The stimulation of dendritic cells may be comparable to surface interactions of macrophages with cytomegalovirus virions, even in the absence of infectious particles. The virions are probably trapped by dendritic cells as immune complexes or through complement receptors. They are then presented to lymphocytes and macrophages as they migrate through the lymphoid tissues, including the spleen. Only 1 in 100 to 400 CD4+ lymphocytes, containing viral DNA, also contained viral RNA. Therefore, macrophages and CD4+ lymphocytes are heavily infected early in the course of the disease and remain latently infected until antigen activation leads to their death as in CD4+ lymphocytes or to further spread of the virus as in macrophages. Furthermore, HIV infection is active in lymphoid tissues while plasma viremia is minimal and peripheral blood mononuclear cells are minimally infected. In advanced-stage disease, patients have been observed to have equal viral burdens in peripheral-blood and lymphoid-tissue mononuclear cells. In early phases of the disease, patients experience lymph node follicular hyperplasia. In intermediate phases of AIDS, the patients have been observed to still have very low levels of virally infected peripheral mononuclear cells. However, in other patients, HIV RNA synthesis was observed to be higher in peripheral blood cells than in peripheral blood cells of patients in earlier phases of the disease, but still lower than in cells of lymphoid tissues (including adenoids and tonsils).[629]

The immunopathology of the disease is relentlessly progressive. In the intermediate stages of the disease, some germinal centers are infected but not others.[629] In late stages, the germinal centers are disrupted by destruction of CD4+ lymphocytes and dendritic cells. In paracortical areas of the nodes, HIV RNA expression is seen in all stages of the disease, but progressively increases as the disease advances. In early stages, follicular dendritic cells efficiently trap virions on their surfaces and present these virions to lymphocytes. Efficiency in trapping virions decreases with disease progression. Therefore, mechanical trapping by follicular dendritic cells and sequestration of infected CD4+ cells in hyperplastic nodes effectively retains the virus in early stages of HIV infection. Antibody complexes with virions further clears virions from the plasma and traps them in the nodes. When these structures are disrupted or destroyed by the infection, plasma viremia results and infection of peripheral blood mononuclear cells ensues.

The mechanism of dendritic cell destruction is unknown, and CD4+ lymphocyte elimination is only partially understood.[380] NO production may be responsible for this immunopathology. Human macrophages and monocytes have been considered incapable of NO synthesis. However, human HL-60 leukemia cells, which are bipotential for the development of monocytic and granulocytic lineages, can be induced to produce NO. Furthermore, human U937 undifferentiated mononuclear leukemia cells latently infected with HIV can be induced to express the virus by treatment with granulocyte/macrophage colony stimulating factor.[358] The chronic, continuous stimulation of follicular dendritic cells with uncleared virus could lead to chronic NO production, which would first stimulate lymphocyte proliferation, when adequate extracellular and intracellular thiols are present, but would ultimately lead to reductant depletion, interaction of NO with oxygen radicals and subsequently, hydroxyl radical formation, resulting in cell death. This latter process could either kill the especially thiol-sensitive CD4+ lymphocytes directly or lead to apoptosis, resulting from cytokine-mediated growth stimulation (IL-1 effects) without reductant growth support. The process of first supporting followed by killing cells in HIV infection is reminiscent of the fratricide of lymphocytes by macrophages in lymph nodes infected with *Mycobacterium leprae* in leprosy.[630,631]

VIII. SUMMARY

The concept of an extracellular disulfide reductase arose with the feeder cell phenomenon in which macrophages provide growth support for primary or cytokine-dependent lymphoid cells. The need for extracellular thiols seemed apparent when 2-mercaptoethanol could be substituted for the macrophages in the feeder layer. However, because thiols are so readily oxidized in extracellular fluid, the 2-ME effect was suggested to provide a mixed disulfide with cysteine, which overcame the transport limitations of lymphocytes for cysteine. This observation led to the conclusion that the extracellular disulfide reductase actually was merely a net apparent effect of amino acid transport. Indeed, the ASC and X_c^- transport systems could provide net import of cystine and export of cysteine in certain cell types such as fibroblasts. However, this entire process is exquisitely sensitive to glutamine and glutamic acid and does not account for overshoot responses to extracellular oxidative stress. A variety of redox and nonredox stressors appear to cause the excessive production of extracellular thiols by macrophages. The process is also induced or inhibited by lipopolysaccharide in a paradoxical fashion dependent on the state of the macrophages. Furthermore, cysteine is a poor reducer of proteins, and thioredoxin and polypeptide hormones such as luteinizing hormone and follicle stimulating hormone are much better reductases/thiol/disulfide isomerases.

Because of the high concentration of intracellular thiols, the two to three orders of magnitude lower concentration of extracellular thiols than intracellular ones, and the presence of most redox-sensitive thiols/disulfides in external cell-surface receptors and intramembrane channels, extracellular reduction of thiols is more critical than thiol import or intracellular reduction. The question remains as to what is the source of the extracellular reducing equivalents necessary to maintain the disulfide reductase activity of thioredoxin or polypeptide hormones bound to their surface receptors. Although reduced glutathione is the principal source of thiols and reducing equivalents from thiols for detoxification and normal cellular growth, it is inadequate as a source of extracellular disulfide reductase because few cells export it as a thiol.

The earliest events in human CD4$^+$ and CD8$^+$ lymphocyte stimulation toward the processes of cellular and humoral immunity are very thiol (GSH) dependent. Oxidative events that rapidly stimulate excessive thiol production are necessary in the early interactions of these cells with mitogens or antigens. Stimulation of lymphocytes is facilitated by follicular dendritic cells and macrophages (both antigen-presenting cells). Dendritic cells never internalize complete HIV virions and become infected; therefore, the metabolic changes induced by HIV in these cells are all cell surface mediated. Chronic stimulation of lymphocytes and virion presentation by dendritic cells and/or macrophages seems an essential part of the immunopathology of AIDS, leading ultimately to the death of the lymphocytes and the dendritic cells and chronic infection of macrophages. Because intracellular GSH is so important to receptor responses of CD4$^+$ and CD8$^+$ lymphocytes and export of cystine is not affected by reductants but import of cystine and growth of the cells is, the existence of a transmembrane thiol transhydrogenase seems not only highly likely but essential.

Chapter 6

NITROGEN OXIDE REDUCTASES

CONTENTS

I. STRUCTURE AND FUNCTION

A. FUNCTIONAL CLASSIFICATION

The purpose of this chapter is to examine the relationships, both structural and functional, among redox enzymes that metabolize nitrogen oxides (N_nO_x), heretofore, exclusively in bacteria, fungi and higher plants. Furthermore, the possibility that they share common features with animal enzymes such as NO synthase and NADPH oxidase and that certain components of all these enzymes are potentially interchangeable will be discussed. The N_nO_x reductases will be reviewed no further than is necessary to accomplish this purpose because they have been thoroughly reviewed elsewhere and these reviews will serve as the necessary background reading for this chapter.[61,632]

Three fundamental questions arise when making structure/function comparisons of nitrogen redox enzymes: (1) What structural features are so fundamental to the reductase/cytochrome functions that they are held by all reductase/cytochromes and, therefore, convey very limited specificity if any? (2) What structural variability is nonessential to function so as to convey the same functions by parallel evolution of tertiary and quaternary structures? (3) What are the fundamental structures essential for redox function and intercomponent interactions (whether achieved by primary and secondary or tertiary and quaternary structures)? Two basic forms of nitrogen oxide reductases (redox chains) exist. They are assimilatory and dissimilatory.[61,632] Nitrate assimilation involves the conversion of nitrate to ammonia by the following series of reactions and incorporation of the ammonia into amino acids or keto acids to form amino acids:

$$NO_3^- + 2e^- + 2H^+ \rightarrow NO_2^- + H_2O \tag{28}$$

$$NO_2^- + 6e^- + 8H^+ \rightarrow NH_4^+ + 2H_2O \tag{29}$$

$$NH_4^+ + \text{glutamic acid} + ATP \rightarrow \text{glutamine} + ADP + P_i \tag{30}$$

$$NH_4^+ + \text{alpha-keto acid} \rightarrow \text{amino acid} \tag{31}$$

Reactions (28) and (29) are catalyzed by assimilatory nitrate and nitrite reductases, respectively. Reaction (30) is catalyzed by glutaminase. Reaction (31) is catalyzed by transaminases and aminases.

Dissimilation involves using nitrate as a final electron acceptor in anaerobic oxidative phosphorylation and further reduction of nitrite to nitrogen gas in denitrification as follows:

$$NO_3^- + 2e^- + 2H^+ \rightarrow NO_2^- + H_2O \tag{32}$$

$$NO_2^- + 2e^- + 2H^+ \rightarrow NO + H_2O \tag{33}$$

$$2NO + 2e^- + 2H^+ \rightarrow N_2O + H_2O \tag{34}$$

$$N_2O + 2e^- + 2H^+ \rightarrow N_2 + H_2O \tag{35}$$

The assimilatory nitrogen oxide reductases are nitrate reductase EC 1.6.6.1–3 and nitrite reductases EC 1.6.6.4 and 1.7.7.1.[632] The assimilatory nitrate reductases are soluble redox enzymes of 200 kDa to 300 kD M_r.[632] The transfer of electrons is from NADPH through flavin adenine dinucleotide (FAD), heme and a molybdenum co-factor to nitrate.[632] Evidence exists for transfer through a sulfhydryl group.[632] Diaphorase activity is also demonstrated by these enzymes.[632] Electron acceptors for diaphorase activity include ferricyanide, cytochrome c and dichlorophenolindophenol. The molybdenum co-factor is not involved in these activities. The reduction of nitrate with reducing equivalents from reduced flavins or viologen dyes is also mediated by the nitrate reductase, bypassing the FAD sites and working directly through the heme or molybdenum co-factor. The diaphorase activity is inhibited by sulfhydryl-binding agents but cyanide inhibits only the molybdenum-mediated activities. Although all assimilatory nitrate reductases are similar, there is structural and catalytic diversity in the redox chain even within a single species (associated with anatomical location of the enzyme).[632] The diversity extends even to the same tissue (e.g., leaves). The majority of leaf nitrate reductases are specific for NADH. The tropical legume *Erythrina senegalensis* has a nitrate reductase that can accept either NADH or NADPH as the initial reductant. Often this bispecific nitrate reductase is found in the same plant tissue as the NADH-specific one (soybean leaves and cotyledons).

In rice seedlings, different induction stimuli results in one nitrate reductase type being formed over the other.[632] Where mutations occur (nar-1a mutant of barley), the bispecific nitrate reductase may be induced to take over the normal function, indicating the involvement of a separate but control-related gene.[633-635]

The assimilatory nitrate reductases have a nonheme-iron-sulfur center and siroheme prosthetic group.[632] Siroheme is an iron tetrahydroporphyrin of the isobacteriochlorin, resembling the porphyrin of chlorophyll. Assimilatory nitrate reductases can be inactivated with NADH or dithionite and reactivated by oxidation by ferricyanide. However, *Neurospora crassa* (fungal) nitrate reductase is not reductively inactivated. *Chlorella vulgaris* (green algae) nitrate reductase requires NADH and cyanide for inactivation. Ammonia indirectly drives nitrate reductase inactivaton in plants. Also, cyanide may be an intermediate in the reverse metabolism of ammonia, leading to inactivation. Either proteases or binding proteins are important in the regulation of nitrate reductases *in vivo*.[632,635]

The nitrite reductases have a FAD-dependent NAD(P)H-diaphorase activity with cytochrome c, ferricyanide, or dichlorophenolindophenol as acceptors and FAD-independent dithionite-nitrite reductase activity (reminiscent of green hemoprotein electron transfer reactions except that GHP uses oxygen as an electron acceptor).[632] A two-electron activity reducing hydroxylamine to ammonia with electrons from NAD(P)H through FAD is also present in assimilatory nitrite reductase.

Genetic control of assimilatory nitrate and nitrite reductase expression is manifested through autogenous means or ammonia metabolites.[632] Small amounts of nitrate reductase are present in cells without nitrate or nitrite.[632] The enzyme is bound to the nirA gene product, preventing induction of niaD and niiA genes for nitrate and nitrite reductase, respectively. The addition of nitrite dissociates the enzyme from the nirA gene product and the nitrate/nitrite reductase genes are expressed.[632] Any structural mutations in nitrate reductase that diminish the nirA interaction lead to constitutive expression of nitrate reductase. Ammonia and glutamine made by glutamine synthetase lead to nitrogen metabolite repression of nitrate/nitrite reductase expression. Glutamine may control nitrate reductase expression by binding to the protein product of *nit-2* (in *N. crassa* and *Aspergillus nidulans*), which is a glutamine- and DNA-binding protein of 22 kD M_r. Glutamine binding would lead to decreased expression of the nitrate reductase.

Glutamine synthetase itself has been proposed as the gene regulator.[632] When there is not enough usable nitrogen in the cell, octameric glutamine synthetase is converted to the tetrameric form, which does not bind nit-2. As a result, nitrate reductase expression is upregulated. Another gene product nmr-1 upregulates nitrate reductase expression by either increasing translation or decreasing the rate of nitrate reductase mRNA degradation.

Dissimilatory N_nO_x reductases are even more diverse than assimilatory ones based on regulation and even cytochrome type.[61] The membrane-bound nitrate reductase of *Escherichia coli* is a three-component system, containing a membrane-imbedded highly labile cytochrome b.[61,636] This lability is based on heme loss and can be partially reversed by reconstitution with hemin following apocytochrome b purification.[61] The operon for this nitrate reductase is designated narGHJI and is coinduced with formate dehydrogenase-*N* (the gene is fdnGHI).[61,637] They are both induced only under anaerobic conditions. The induction requires the activator protein Fnr. The nar and fdn operons are further induced by nitrate under anaerobic conditions.[61,637-639] This latter process is mediated by the activator protein NarL. Integration host factor (IHF), a heterodimer encoded by himA and himD (hip) genes, produces a loop in the DNA that brings bound NarL in proximity to the transcription initiation site.[639] This idea has recently been challenged with the suggestion that the intervening sequences actually specify a structure that allows for direct interaction between the upstream activator NarL and this transcription complex for nitrate reductase.[61,637,638]

Nitrate inhibits synthesis of other anaerobic respiratory enzymes such as fumarate reductase and synthesis of fermentative enzymes. Another regulatory NarX gene produces a product that mediates nitrate repression of the fumarate reductase gene frdABCD.[61] The NarX product may be a sensor that modifies NarL activity. Synthesis of the *E. coli* dissimilatory nitrate reductase is repressed by oxygen but not inhibited by ammonia.[61] Another nitrate reductase found in *E. coli* that is constitutively expressed in low amounts is termed nitrate reductase Z.[640]

The Z enzyme is also a molybdenum-containing enzyme and has chlorate reductase activity like the inducible enzyme.[640] This activity is lethal because the chlorate is converted to chlorite, which can dismutate or be further reduced by other enzymes to chlorinating hypochlorous acid. True chlorate reductase C shows inhibition of its induction by nitrate, unlike nitrate reductase Z.[640] Generally, chlorate inhibits assimilatory nitrate reductase and is a substrate for dissimilatory nitrate reductases.[61,632] The Z enzyme is believed to couple formate oxidation to nitrate reduction through quinones and type-b cytochromes.[640] The Z enzyme is not induced by nitrate, repressed by oxygen, or activated by the fnr protein product.[640] It may be that enzyme Z acts as a regulatory protein for the inducible reductase.[640]

The oxidative Gram-negative obligate aerobic bacteria *Alcaligenes eutrophus* (strain H 16) also contains three distinct nitrate reductases.[641] One is assimilatory, and the other two are dissimilatory. The assimilatory one is repressed by ammonia and insensitive to oxygen. The membrane-bound dissimilatory nitrate reductase, as expected, is formed in the absence of oxygen and is insensitive to ammonia repression. The third nitrate reductase is a soluble periplasmic enzyme that is insensitive to both ammonia and oxygen (much like the enzyme Z of *E. coli*). The periplasmic nitrate reductase (NAP) is encoded by a megaplasmid pHG1 and is constitutive (insensitive to nitrate), unlike the other two nitrate reductases. The enzyme has two structural genes napA and napB, encoding for proteins of 93 and 18.9 kDa, respectively. The K_m for nitrate for NAP is 0.12 mM. The physiologic electron donor for this enzyme is unknown. Other microorganisms with soluble periplasmic nitrate reductases include *Rhodobacter capsulatus, R. sphaeroides* f. sp. *denitrificans,* and *Thiosphaera pantotropha.*[641]

These periplasmic nitrate reductases are typically composed of type-c cytochromes and a molybdenum-containing subunit.[641] *Thiosphaera pantotropha* is capable of aerobic denitrification, but *A. eutrophus* is not. NAP activity is very low under anaerobiosis. Also, partially purified NAP can only use formate as an electron donor. The enzyme has been proposed as being important for transition from aerobic to anaerobic growth, since mutants lacking it show delayed growth under anaerobic conditions.[641]

The remaining steps in the denitrification process are catalyzed by the enzymes nitrite reductase, nitric oxide reductase, and nitrous oxide reductase, respectively.[61,642-644] These have been most intensely investigated in *Pseudomonas stutzeri* and *Ps. aeruginosa.*[643-649] The nitrite reductases (ferrocytochrome c_{351}:oxygen oxidoreductase; EC 1.9.3.2) of these microorganisms is a cytochrome cd_1 complex.[647,648] Cytochrome d_1 (previously designated a_2) has the same heme type as the terminal oxidase of bacteria in which the ring A vinyl side chain is replaced with a secondary ethanol group. The structural gene for nitrite reductase is nirS.[647] In *Ps. stutzeri,* the gene is closely located to the nitrous oxide reductase gene nosZ.[644] The two genes are separated by 14 kb. Several genes for copper handling are located between the two genes, which is not unusual since nitrous oxide (N_2O) reductase is a copper-containing enzyme.[642] A gene cluster of 30 kb encompasses the nir (nitrate reductase), nor (nitric oxide reductase) and nos (nitrous oxide reductase) genes.[649] The subcluster of genes are arranged in the order nos-nir-nor.

Two kinds of dissimilatory nitrite reductase exist in bacteria, but no instance of both enzymes existing in the same species is known.[61] The *Ps. stutzeri* nitrite reductase is a tetraheme containing cytochrome (EC 1.9.3.2), and the alternate nitrite reductase is a copper-containing enzyme (EC 1.7.2.1) (as in *Achromobacter cycloclastes*).[61] The former one is more widely disseminated among species of denitrifying bacteria.

Pseudomonas aeruginosa, A. cycloclastes, and *Ps. stutzeri* nitrite reductases have been used to show the formation of nitrosonium ions and to show that the formation of nitric oxide is the major pathway of nitrite reduction in denitrifying bacteria.[650-652] NO interactions on the enzyme leads to another minor pathway for producing nitrous oxide. Furthermore, organic nitrosations (such as that of aromatics) can be catalyzed as a minor pathway by these nitrite reductases.[651] For nitrosation of aniline by nitrite catalyzed by the *Ps. aeruginosa* enzyme, the $k_{cat}/K_m \times K_m' = 5.8 \times 10^{10}$ M^{-2}s^{-1} at pH 7.5. The $k_{cat} = 0.077$ s^{-1}, $K_m(HNO_2) = 3.34 \times 10^{-10}$ M, K_m (nitrite) = 5.3×10^{-6} M and K_m' aniline) = 4 $\times 10^{-3}$ M. The nonenzymatic acid-catalyzed nitrosation of *o*-chloroaniline has a third-order rate constant of 175 M^{-2}s^{-1}.[650,651]

In *Ps. stutzeri* (strain Zo Bell), a gene product of gene nirQ is a transcriptional activator for nirS, the structural gene for nitrite reductase.[644] The activator is a protein of $M_r = 30,544$. Both nitrite and nitric oxide reductase activities are decreased with mutations of nirQ, but both enzymes are still transcribed and translated into proteins. NO is a toxic product. Therefore, the loss of nitric oxide reductase activity leads to a conditional lethal mutant.[645] The normal free NO in steady-state denitrification by *Ps. stutzeri* (strain Zo Bell) has been estimated to be 30 nM. In other bacteria, it may be as low as 1 nM. In the intergenic, nontranscribed/nontranslated region of nirQ and nirS (the entire operon for nitrite reductase is nirSTBM), there are two Fnr-binding sites, indicating its potential regulation of these genes.[644,649] Therefore, nirQ protein would require denitrifying conditions for expression.

The nitric oxide reductase is a 17-kD cytochrome c combined with a 38-kD membrane-bound cytochrome b subunit.[645] The K_m of this enzyme for NO has been reported to be 60 μM with a pH optimum of 4.8. Again, the synthesis of this cytochrome bc enzyme is strictly dependent on anaerobic culture conditions. A similar cytochrome bc enzyme has been isolated from *Paracoccus denitrificans*. Besides keeping the steady-state concentration of NO low with nitric oxide reductase, the binding of NO to cytochrome synthesized for that purpose during denitrification may be another mechanism of decreasing toxicity. An example of NO-cytochrome c' complex formation has been noted.[645]

The final enzyme in the denitrifying sequence for consideration is nitrous oxide reductase. The entire operon for nitrous oxide reductase (EC 1.7.99.6) is nosZDFY, which has been sequenced in *Ps. stutzeri*.[61,642,644] This nitrous oxide reductase is a multi-copper-containing enzyme.[642] It is a membrane protein, on the periplasmic side of the membrane, with an $M_r = 120–140$ kD.[61] The copper is bound in mixed valence form (reminiscent of galactose oxidase). The nosDFY genes are involved with copper processing for the reductase.[642] The protein NOSA from the nosA gene is an outer-membrane protein that forms channels for copper transport to support transfer of copper ions to the reductase.[642] It has been suggested that the release of copper ion from NOSA is a chemical-energy-requiring step (ATP or GTP).[642] The nosA gene is repressible by copper. The other copper-processing genes of the operon for nitrous oxide reductase are nosD, a periplasmic protein; nosF, a cytoplasmic nucleotide-binding protein; and nosY, an inner-membrane protein.[642]

B. STRUCTURAL SIMILARITIES TO OTHER REDOX ENZYMES

From the perspective of this book, the two families of nitrogen oxide reductases that are most important because of their cytochrome b content are nitrate reductase and nitric oxide reductase. The nitrite and nitrous oxide reductases, positioned, respectively, between and after the other two nitrogen oxide reductases in the nitrogen oxide metabolizing pathway, more closely resemble terminal copper and/or cytochrome oxidases involved in mitochondrial respiration of eukaryotic organisms.

The nitrate and nitric oxide reductases bear strong similarities to cytochrome b reductases, NO synthases, and NADPH oxidase.[653] All these bear the common properties of hemoproteins that interact with quinones and some that are membrane-bound type-b cytochromes or appear to have had a membrane-bound precursor (ontologically or phylogenetically).[58]

First, some common features of type-b cytochromes must be revisited to understand the functional significance of certain structural features. The polypeptides that bind the protoheme (protoporphyrin IX-type) in membrane-bound cytochrome b enzymes, cross the lipid bilayer of the membrane several times (at least three) and have the metal porphyrin embedded in the lipid.[58] This feature allows interactions with quinols and quinones that diffuse through the membrane lipids.[58] The transmembrane characteristic is also essential for binding other nonintegral redox proteins to the membrane.

There is high variability in the primary structure of these cytochromes.[58] However, they do possess common features associated with quinone/semiquinone binding. The iron ions of their hemes also show strained, bisimidazole coordination, which may contribute largely to their instability (and function).[58] The variability in the optimal maxima of visible light absorption of the reduced alpha band of the heme of the cytochromes gives no discrete means of identifying the cytochromes or relating to the specific function of a given type-b cytochrome.[58] However, based on electron spin studies, the spectroscopic property of high anisotropic low-spin signal (HALS signal) has been found to be a common feature of quinonoid-binding cytochrome b's.[58] The M_r of most cytochrome b's or b-like subunits is around 20 kD. Only the histidines that coordinate with the iron in the hemes are highly conserved in the general group of quinonoid-binding type-b cytochromes. However, no methionine is coordinated with the iron in these cytochromes. The diversity in amino acid sequence of type-b cytochromes makes the process of deducing structure/function and evolutionary relationships very difficult.[58]

Methods for examining and comparing these relationships have included a histidine index examining peptide relationships around the iron-coordinating histidines and examining specifically enriched residues (nonrandom collections) around ligand binding sites.[58] These techniques have drawn together type-b cytochromes in structure and function into relationships that would have been overlooked otherwise because of overall nonhomologies.

The ubiquinol or menaquinol:nitrate reductase redox chain of *E. coli* has a type-b cytochrome, b-556, which is encoded by the narI gene.[58,637] This cytochrome has four transmembrane segments. Four of its histidines are available for coordination. Therefore, it can coordinate two b-hemes (protohemes). This has been confirmed in *E. coli* and *Ps. denitrificans* nitrate reductases.

The *E. coli* membrane-bound nitrate reductase is composed of three subunits: (1) A (alpha) with an M_r = 145 kD; (2) B (beta) with an M_r = 60 kD; and (3) C (gamma) with an M_r = 20 kD.[61,636] Subunit C is the type-b cytochrome (its structural gene is narI). The subunit composition of the complete nitrate reductase is usually 2A:2B:4C. Another form (soluble) has a 2A:2B composition. The entire complex has 28 ± 2 atoms of iron and 1.35 atoms of molybdenum. When C was not included, the iron content was 24 ± 2 atoms of iron and 1.18 atoms of molybdenum per molecule of enzyme. Subunit C is easily denatured but can be partially recovered by dialysis against hemin. In mutants lacking subunit C, there is an overproduction of subunits A and B, showing that C has feedback control on their synthesis. Like green hemoproteins, subunit C aggregates upon brief boiling or storage under refrigeration, and treatment with dithionite causes bleaching (or loss) of the heme spectra and aggregation, also. The assemblage of the three subunits in the cytoplasmic membrane of *E. coli* involves subunit A's being accessible on the inside of the cytoplasmic membrane, subunit B's being buried within the membrane altogether and subunit C's being accessible to the outside of the cytoplasmic membrane. This overall organization is reminiscent of NADPH oxidase structure.

In general, NADH, succinate, lactate, formate, glyceraldehyde, and hydrogen have been demonstrated to serve as sources of reducing equivalents for dissimilatory membrane-bound nitrate reductases (type A; EC 1.7.99.4).[61] These reductants are linked by membrane-bound dehydrogenases to nitrate reductase through a quinone and cytochrome. NADH and formate are the preferred substrates for bacteria when carbohydrates are available in culture media. That formate would be a preferred reductant is not unexpected since its E_0' (mV) is –432, which makes it a more powerful reductant than hydrogen (–414 mV), ferredoxin (–398 mV), or NADH (–320 mV). However, when *E. coli* is restricted to glycerol, succinate, or lactate as an electron donor with nitrate as the acceptor, then the cytoplasmic membranes contain succinate, glycerol phosphate, and lactate dehydrogenases.

The formate dehydrogenase of *E. coli* is a molybdoprotein containing selenium (M_r = 590 kD).[61] The electron acceptor is cytochrome b_{FDH}. It transfers its electrons to the cytochrome b of nitrate reductase through a ubiquinone. The molybdenum-iron-sulfur protein component of the nitrate reductase then interacts directly with the nitrate, reducing it to nitrite. Nitrate reductase A's can do more than reduce chlorate, as mentioned earlier; they can also reduce bromate.[61] They are very sensitive to azide inhibition. They are also inhibited by cyanide, but not by carbon monoxide. The *E. coli* enzyme prefers ubiquinone over menaquinone as a mediator of electron transfer. Typical of all quinonoidal interacting type-b cytochromes, 2-alkyl-*N*-hydroxyquinolines, such as 2-*n*-heptyl-4-hydroxy-quinoline-*N*-oxide (HQNO), inhibit nitrate reductase A activity.[58] Gram-positive bacteria, which generally lack ubiquinones, depend on naphthoquinone as a substitute in the nitrate reductase.[58,61]

All four steps of nitrate reduction to nitrogen gas can be considered potentially capable of oxidative phosphorylation. In vesicle-cell-free systems of *E. coli, Micrococcus denitrificans,* and *Ps. aeruginosa,* nitrate reduction drove phosphorylation.[61] Dinitrophenol uncoupled these systems, decreasing phosphorylation and increasing the rate of electron transfer. Phosphate-per-two-electron transport ratios of 0.9 for NADH, 0.65 for glutamate, and 1.1 for citrate were observed with nitrate reduction. However, based on proton transloca-tion, the membrane-bound components of the denitrification process, nitrate reductase and nitric oxide reductase, should be the most likely cytochromes to directly convert electron transfer to phosphorylations by a membrane-associ-ated ATPase driven by proton-motive force.

The assimilatory nitrate reductases (type B) follow a different pattern.[632] Collectively, they are soluble enzymes, they cannot reduce chlorate to chlorite (show inhibition), and they are relatively insensitive to azide inhibition. The degree of monovalent binding of FAD to assimilatory nitrate reductases is variable, with *N. crassa* showing weak binding and others strong binding.[74,632] There is a key sulfhydryl group that mediates electron transfers between the NADH and FAD. The reduced FAD transfers electrons to cytochrome b_{557} (alpha peak). The cytochrome transfers the electrons to the molybdenum bound to a low-molecular-weight co-factor, urothione or molybdopterin. The molyb-denum co-factor then reduces the nitrate to nitrite.

Fungal nitrate reductases (except for that of *As. nidulans,* which has sub-units of 59 kD and 138 kD) are homodimers with an $M_r > 200,000$, specific for NADPH, with a pH optimum of 7.5 and an easily lost FAD.[632] The specific activities of these enzymes are about 100 μmol nitrate reduced per minute per milligram of enzyme.

Algal nitrate reductases are represented by the green alga (*C. vulgaris*), which has been the most studied.[632,654] The nitrate reductase is a heterodimer of 360 kD. These enzymes tightly bind FAD.

In general, most higher plant nitrate reductases are highly variable.[632] Most leaf nitrate reductases are specific for NADH and are homodimers of subunits of about 115 kD.[632,655]

Assimilatory nitrite reductases have an iron-sulfur center and siroheme groups.[632] The iron-sulfur groups are organized as tetranuclear clusters of four irons and four sulfurs, like the nonheme iron proteins of mitochondria and endonuclease III. The siroheme is also found in sulfite reductases. The siroheme provides the most probable binding site for the nitrite, and the iron-sulfur centers are chemically interactive with the sirohemes. The assimilatory nitrite reductase is a 61-kD protein (in spinach leaves) and uses ferredoxin as the electron source (typical of photosynthetic plants).[632]

In the fungus *N. crassa*, the nitrite reductase is a 290 kD homodimer flavoprotein, which uses either NADPH or NADH as the electron donor.[74,632] Both the higher plant and fungal nitrite reductases are inhibited by the sulfhy-dryl reagent *p*-hydroxymercuribenzoate (like assimilatory nitrate reductases), by cyanide, sulfite, and carbon monoxide.[74,632]

The nitrate reductases of plants resemble some mammalian type-b cyto-chromes. In the green alga *Chlamydomonas reinhardtii*, the nitrate reductase gene generates a transcript of 3.4 kb with a high degree of sequence homology with the sequences for heme- and FAD/NADH-binding domains of that of the tobacco plant and *Arabidopsis thaliana* nitrate reductases and human NADH cytochrome b_5 reductase.[653,656,657] In the tobacco plant (*Nicotiana plumbaghrifolia*), the nitrate reductase heme domain is very similar in structure to cytochrome b_5.[658] The higher plant nitrate reductase (EC 1.6.6.1) resembles the soluble cytochrome b_5 reductase and the cytochrome b_5 of erythrocytes in structure and function. The nitrate reductase shows homology to cytochrome b_5 near its mid-chain amino acid sequences and homology to cytochrome b_5 reductase (EC 1.6.2.2) at the C-terminal 260 amino acids of the nitrate reductase.[653] In the N-terminal region, nitrate reductase resembles the molybdopterin region of chicken liver sulfite oxidase (EC 1.8.2.1).[58,632,653]

The general organization of assimilatory nitrate reductases is linear, with the molybdopterin-binding domain at the N-terminal, the cytochrome b domain in the middle, and the FAD domain at the C-terminal.[632] It has been suggested that the assimilatory nitrate reductases belong to a family of flavoprotein pyridine nucleotide cytochrome reductases in which 14 amino acid residues in the FAD domains are invariant.[659] Other members of this family are NADH:cytochrome b_5 reductase, NADPH:nitrate reductase, ferredoxin:NADP$^+$ reductase, NADPH:cytochrome P450 reductases, NADPH:sulfite reductases, and *Bacillus megaterium* cytochrome P450 BM-3.[659] This family may be extended to include relationships with the membrane component of NADPH oxidase (gp91-phox and p22-phox) and nitric oxide synthase (EC 1.14.23), both constitutive and inducible.[653]

Cytochrome b_{558}, the membrane component of NADPH oxidase, has three distinct domains: (1) the N-terminal of its gp91-phox, containing the binding site for p22-phox (or its heme); (2) the FAD-binding domain in the middle of gp91-phox; and (3) the NADPH-binding domain at the C-terminal.[653] The FAD-binding and NADPH-binding regions have homology with the cytochrome b_5 reductases, glutathione reductases, inducible and constitutive NOS, nitrate reductases (fungal), cytochrome P450 reductases, and ferredoxin reductases.[653] NOS has strong homology to cytochrome P450 reductase and is a type-b P450 cytochrome as stated earlier. Therefore, both NOS and NADPH oxidase appear to be related flavocytochromes.

The FAD-binding site homology of gp91-phox to glutathione reductase and the NADPH-binding site homology of gp91-phox to ferredoxin reductase have been interpreted as possible indications of cytochrome b_{558} of NADPH oxidase arising from a fusion of ancestral genes from two distinct groups of flavoproteins.[653] To carry this evolutionary perspective further, the possibility that quinone-reacting hemoproteins evolved from a common ancestor has been considered.[58]

This hypothesis has been proposed by Esposti.[58] His evidence for this hypothesis is as follows:

1. Either the first or second transmembrane helix of these cytochromes contains a heme-binding site on the negative side of the membrane.
2. The N-terminal side of the first helix contains a hydrophobic loop rich in aromatic amino acid residues, especially tryptophan.
3. The first transmembrane helix starts with glycine or proline or serine, which is the conserved glycine 33 in yeast mitochondrial cytochrome b.
4. At the third turn of the first helix toward the interior of the membrane is located a histidine (histidine 56 in cytochrome b_{556} of nitrate reductase that corresponds to glutamine 43 of mitochondrial cytochrome b and glutamine 18 in *E. coli* dimethylsulfoxide reductase).
5. At the fourth turn into the lipid of the membrane is located a glycine that corresponds to glycine 47 in mitochondrial cytochrome b (mutations in the region lead to antimycin resistance).

These local conservations suggest that quinonoid-interacting cytochrome b's may have come from a common ancestor. The diversity from the ancient common ancestor may have evolved by multiplication, diversification, or fusion as noted above for NADPH oxidase.[58,653]

Cytochromes b_{561} (a diheme) and b_{556} of succinate dehydrogenase subunit C (SDHC) of *E. coli* are the simplest of quinonoid-interacting type-b cytochromes and have only three transmembrane segments.[58] Triplication and fusion of the b cytochrome of succinate:Q reductase may have led to the b cytochrome of mitochondrial bc_1 complex. A similar occurrence could have resulted in nitric oxide reductase.

In the mitochondrial cytochrome b, there are nine hydrophobic regions (like the original three transmembrane regions).[58] The genes of *E. coli* subunits C and D of succinate dehydrogenase overlap and could have fused to produce the hydrophobic regions of mitochondrial cytochrome b. Cytochrome b_6 of the intersystem photosynthetic pathway in the b_6f complex is associated with a protein that is homologous to the C-terminal region of mitochondrial b cytochrome, which makes up three alpha helices and has a hydrophobic amino acid distribution that is almost identical to that of *E. coli* succinate dehydrogenase subunit C.[58]

The third type of nitrate reductase, NAP (soluble periplasmic type), is composed of a small cytochrome c subunit (13-kD cytochrome C_{552} in *R. capsulatus*) and a larger molybdenum-containing subunit (83 kD in *R. capsulatus*).[641] In *Al. eutrophus*, the cytochrome c has two heme-binding sites. This type of nitrate reductase is probably a result of either parallel evolution, yielding the same function as the type-b cytochrome-containing enzymes but from a different origin, or the substitution of a more stable cytochrome c for b.

The *Ps. stutzeri* cytochrome cd_1 is another nitrogen oxide reductase that interacts with a cytochrome b system, but is not one. The nir and nos genes required by *Ps. stutzeri* for nitrite and nitrous oxide reduction, respectively, are closely linked and controlled.[644] NirD is involved in the heme d_1 cytochrome synthesis; nirS is the structural gene for nitrite reductase. The nirS region is composed of genes nirSTBM.[647] The gene order of *Ps. stutzeri* is quite different from that of *Ps. aeruginosa* in respect to cytochrome cd_1 genes and cytochrome c_{551}, even though these organisms are closely related.[648,649] The nirS gene is followed by nirM and open-reading frame 5 in *Pseudomonas aeruginosa*. Beyond the 3′ end of nirM in *Ps. aeruginosa*, there are 232 bp that are not translated.

In *Ps. stutzeri*, nirS and nirM-open-reading frame 5 are separated by nirT and nirB.[649] The nirT protein product is a tetraheme that transfers electrons to cytochrome cd_1.[647] The N-terminal of the nirT protein is hydrophobic and could anchor it to the cell membrane for electron transfer to a periplasmic electron acceptor. The acceptor could be cytochrome c_{551} or the reductase directly. NirT is connected to the nirB gene, which codes for cytochrome c_{552} and is not translated except under anaerobic conditions. This latter cytochrome can form a peroxidase if partially proteolyzed. It may be that cytochrome c_{552} is the important electron transfer intermediary rather than cytochrome c_{551}. The latter is not regulated by anaerobic conditions, and the gene nirM for c_{551} is 320 bp from nirB. Therefore, it probably belongs to a different operon from cytochrome cd_1.

C. AS GROWTH REGULATORS

At first glance, the growth-regulating functions are obvious for nitrogen oxide reductases, in that they provide ammonia for amino acid synthesis and ATP as useful chemical energy in the cell for all manner of synthetic and catabolic processes. In higher plants, other biochemical changes indicate that the nitrogen oxide reductases participate in more sophisticated metabolic control mechanisms than this first-level control. In barley plants, nitrate reductase synthesis depends on both nitrate induction and light, both being positive coinducers.[660] Also, in barley, oxidative phosphorylation uncouplers increased nitrate reduction, as one might expect, in both light and dark, and photosynthetic inhibitors had no effect.[660] However, more nitrite (which is usually at low steady-state concentrations in barley leaves) accumulated in leaves treated with photosynthetic inhibitors (atrazine or 3-(3′,4′-dichlorophenyl)-1,1-dimethylurea) whether in the dark or the light. It is possible that these inhibitors may have affected nitrite reductase directly or the availability of substrate (nitrite transport). Nitrite assimilation showed greater sensitivity to oxidative phosphorylation uncouplers in the dark than in the light. In the dark, >74% of the nitrate was reduced to nitrite and further reduced whereas >95% of the nitrite was further reduced in light.

Other evidence of control is shown in the hysteretic behavior of squash (*Cucurbita maxima* L. cv. *Buttercup*) nitrate reductase.[661] This enzyme shows slow allosteric activation by reduced nucleotides, either NADH or NADPH, even though it can only use NADH as a substrate for reduction.[661] This enzyme is inherently unstable, undergoing a 50% decrease in activity in 2 hours at 25°C after purification. However, about 80% of the lost activity of the enzyme has been restored in 2 min when treated with 200 μM of NADH or NADPH. The exact function of this hysteresis (slow reactivation process) is unknown. However, other enzymes central to metabolic (carbohydrate metabolism) control, such as glyceraldehyde-3-phosphate dehydrogenase, phosphofructokinase, and phosphoenolpyruvate carboxylase, are all hysteretic enzymes in plants. Another obvious growth-regulating function is that of nitric oxide (or nitrite to a lesser degree) production, which as noted earlier can be toxic to bacteria if allowed to accumulate. It can deaminate nucleotides, inhibiting both DNA replication and RNA synthesis, and therefore, protein synthesis. It can also potentially trigger gene expression.

D. AS DIAZOLUMINOMELANIN SYNTHASES

As pointed out in Chapter 2, diazoluminomelanin (DALM) is a photoactive polymer that interacts with type-b cytochromes (green hemoproteins). It also appears to interact with a membrane-bound cytochrome component (putatively the cytochrome b of NADPH oxidase) in HL-60 human promyelocytic leukemia cells.[122] Because DALM is made by organic synthesis using codiazotization of luminol and 3-amino-L-tyrosine with acidified nitrite, the manufacture of DALM by bacteria with active nitrate and nitrite reduction seemed plausible.[662,663] The nitrite produced by nitrate reduction under acidic conditions should have produced DALM in growth medium containing luminol, nitrate, and 3-amino-L-tyrosine (3AT). Furthermore, bacteria that reduce nitrite to nitric oxide should have been able to form DALM in neutral to alkaline growth medium. The results were surprising when several different species of bacteria with nitrate and nitrite reductases were compared.[662] In general, when the concentrations of 3AT were varied (0.2 to 4 mM), even though *E. coli* varied in growth and pigment (DALM) production, concentrations equal to or greater than 0.4 mM inhibited *E. coli* growth. Microaerophilic (candle jar, carbon dioxide atmosphere) or stab cultures (near anaerobic growth) were required for the best DALM production by *E. coli*. These conditions are in agreement with the anaerobic induction of the nitrate reductase in *E. coli* (except for the minor Z enzyme). *Bacillus cereus* and *B. anthracis* (Sterne strain) grew on the medium up to a concentration of 4 mM 3AT (the latter showed significant inhibition of growth). The best growth and DALM production by these species were at 2 mM 3AT. Unlike the *E. coli* strains used, growth in carbon dioxide or in stab agar cultures inhibited DALM production (but not necessarily growth) of the *Bacillus* spp.[663] DALM production by these bacilli was best in air on petri dishes containing the 3AT/luminol/nitrate agar.

Pseudomonas aeruginosa also grows well and produces pigment on 3AT/luminol/nitrate agar like *Bacillus* spp. However, it did not produce the comparable luminescence when activated with sodium bicarbonate and hydrogen peroxide solution and heat.[664] *Pseudomonas stutzeri* also produced results

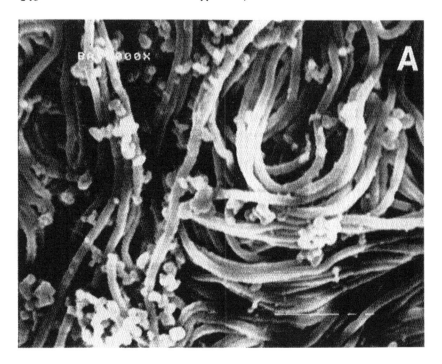

FIGURES 20A and B.

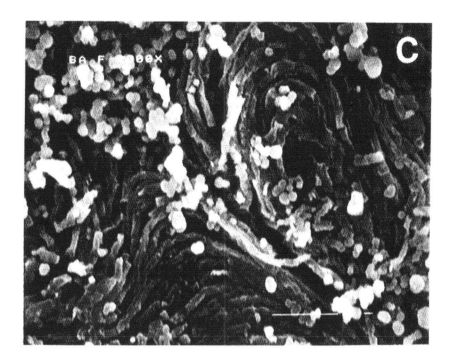

FIGURE 20. Sequence of scanning electromicrographs of *Bacillus anthracis* (Sterne strain) grown on 3-amino-L-tyrosine/luminol/nitrate agar medium. (A) After 48 hours of culture on control tripticase soy agar, showing Medusa head formation with spores present; (B) after 48 hours of growth on 3AT/luminol/nitrate agar, showing bending and curling of bacilli without the extreme elongation into Medusa strands; (C) 6 days of culture on 3AT/luminol/nitrate agar, showing twisting and shrinking of bacilli. (Magnification × 2000.)

similar to *Ps. aeruginosa*, except when a nitrogen gas negative mutant was used. In this latter case, the *Ps. stutzeri* produced abundant thermochemi–luminescence. These results are explainable based on the pseudomonads conversion of nitrate to nitrogen gas and the inability of the 3AT to efficiently trap the intermediates that bring about diazotization, except in the case of the *Ps. stutzeri* mutant.

When the DNA of *B. anthracis* (Sterne strain), *B. cereus*, *Ps. aeruginosa*, and *Ps. stutzeri* were probed by Southern blotting with a ^{32}P-labeled gene probe containing the 1.1-kb piece of the 3.6-kb nitrate reductase gene of barley, they demonstrated hybridization under stringent conditions.[664] *Escherichia coli* strains probed in this way showed no hybridization. Therefore, it was concluded that the former aerobic bacteria contain nitrate reductases similar to the assimilatory nitrate reductase of higher plants.

Figure 20 (A, B, C) shows scanning electronmicrographs of *B. anthracis* undergoing progressive morphological changes on the 3AT/luminol/nitrate medium. These results are believed to be caused by cross-linking attachment of the

DALM polymer to the membranes of growing bacteria, restricting normal growth of the bacteria. The result of the resistance created by the attached polymer against elongation of the microorganism is the observed coiling and twisting of the bacilli.[665] The string-of-pearl formation of these bacteria from penicillin treatment during growth (from failure to form cell-wall components) is accelerated and truncated on 3AT/luminol/nitrate medium.[663] Although the "pearls" form on this medium in 3 hours rather than the usual 6–18 hours on other media, they do not proceed to bursting, but become fixed (from cross-linking).

When the *B. anthracis* was grown on 3AT/luminol/nitrate medium for 24 hours, removed, washed, and resuspended in physiologic saline containing 0.25% sodium bicarbonate and 0.03% hydrogen peroxide and exposed to 2450 MHz (100 mW/g) microwave radiation for 30 min at 37°C, greater than 97% of the bacteria were killed.[666] If the same bacteria were grown on trypticase soy agar and subjected to the same chemical and exposure conditions, there was no significant killing, that is decrease in colony plate counts after exposure. The DALM on one hand accelerated growth and on the other sensitized the cells to oxidative killing facilitated by microwave radiation.

Attempts have failed to produce DALM in RAW 264.7 cells by growing them in medium containing 3AT and luminol and stimulating their NO synthase activity with gamma-interferon and LPS.[218] When HL-60 cells were grown in 3AT and luminol medium, they produced aminomelanin, which lacked the properties of DALM.[419] The aminomelanin production of HL-60 cells seemed to be associated with their myeloperoxidase-containing granules. However, when these aminomelanin-laden cells were treated with nitrite at pH 4.0, they developed large disrupting fibers of DALM that bore the same photochemical and microwave-absorbing properties of synthetic DALM. Part of the problem in DALM production by NO synthase in eukaryotic cells may lie in the production of alpha-amine-derived diazonium groups on the 3-amino-L-tyrosine at neutral to alkaline conditions in the cells. These alpha-diazonium groups are unstable and rapidly break down into nitrogen gas and a mixture of alcohol and alkene derivatives of 3-amino-L-tyrosine. The absence of alpha-amino groups should prevent polymerization of 3-amino-L-tyrosine (see Chapter 2). Also, thiols produced by these macrophage-like cells could displace the aromatic diazonium groups by nucleophilic displacement, resulting in inhibition of the polymerization process. Another possibility that could prevent DALM formation in animal cells is azoreductase activity, which would destroy the diazonium groups.[667]

Human, guinea pig, and mouse skin cells (epidermal cells) have shown extensive azoreductase activity in both the cytosolic and microsomal fractions of the cells.[667] They readily reduce various azo dyes (Sudan I and FD&C Yellow No. 6) to their corresponding amino aromatics. Furthermore, organic nitroreductase, azo reductase, dechlorinase, and dehydrochlorinase activities have been found in the mucosa of the small intestines of rats (excluding the interference of intestinal microflora).[668]

II. INTERACTIONS WITH NITRIC OXIDE SYNTHASE

In an effort to produce DALM in animal cells, the expression of a plant aerobic nitrate reductase was attempted in a variety of mouse and human cells.[521] A 1.1- or 800-kb fragment of the 3.6-kb gene of assimilatory nitrate reductase of barley plant was inserted in the the pSV_2neo plasmid. The plasmid was then transfected into target cells by electroporation (450 V, 25 μF, 5 pulses). Cells were selected for resistance and plasmid content by growth in medium containing 500 $\mu g/ml$ of neomycin. Three clones of RAW cells were produced: (1) NR 10_1, (2) NR10_2, and (3) N800$_5$. The cell types in which transfection was attempted were RAW 264.7 (murine mono-cyte/macrophage), P388D$_1$ (murine macrophage), J774A.1 (murine macrophage), EL-4 (murine thymoma), HL-60 (human promyelocytic leukemia), U937 (human monoblastic leukemia), K562 (human erythroblasts), B16 (murine melanoma), HeLa (human cervical carcinoma), and NIH 3T3 (murine fibroblasts). Nitrate reductase activity was detected only in RAW 264.7, P388D$_1$, J774.A1, and EL-4 cells. The best results were seen with RAW 264.7 cells (NR10$_2$ clones) primed with gamma-interferon and LPS. The interactions with these stimuli and the incomplete gene fragment expression suggested that the nitrate reductase activity was integrated with the NOS activity. When the RAW 264.7 cells (transfected and nontransfected) were examined in greater detail, effects on growth and spontaneous nitrite formation (no nitrate added) of transfected cells vs. nontransfected cells were remarkable. First, Figure 21 shows that the transfected cells grew more slowly and retained higher viability with days in culture. The transfectants produced more nitrite (as shown in Figure 21 by the production of accumulated color change with the Greiss reagent). On a per-million-cell basis, in the absence of added nitrate, the nitrite production peaked in the transfected cultures on day 2 and declined thereafter to 2 μM by day 7 (Figure 22). When nitrate was added to the medium, the nitrite production by transfected clones rose nearly exponentially (Figure 22) between days 1 and 3 and more slowly thereafter. The control (nontransfected) nitrite production also rose with days in culture, but much more slowly (Figure 22). This change in controls could be attributed to weak endogenous nitrate reductase activity or spontaneous NO synthase activity.[521] Nontransfected P388D$_1$ cells treated with nitrate (39.5 mM) showed increased TNF production (see Table 2), which could have, in turn, stimulated corresponding increases in NO synthase activity.[669] However, transfected P388D$_1$ cells showed no such increase in TNF production when nitrate was added to their medium, although they reduced nitrate to nitrite (shown by the Greiss reaction). Also, inhibition studies and previous reports (see Chapter 5) have shown that a constitutive NOS in RAW cells could have provided the nitrate substrate for the transfected cells in the absence of added nitrate.

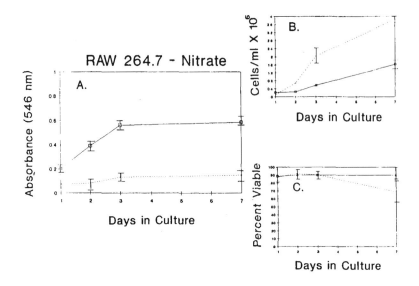

FIGURE 21. Total nitrite production (A), cell number (B), and viability (C) of RAW 264.7 cells transfected with a 1.1-kb fragment of the barley nitrate reductase gene (solid line). The 546-nm absorbance is of the diazotized Greiss reagent, indicating nitrite production; trypan blue was used to determine cell viability. The nontransfectants (dashed line) constitute the parent cell line from which the transfectants were derived by electroporation and plasmid introduction. The error bars are for triplicate samples. See the text for further details.

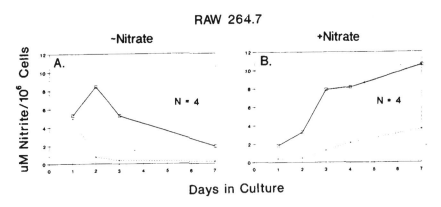

FIGURE 22. Comparison of specific nitrite production of nitrate-reductase transfected (solid line) with nontransfected RAW 264.7 (dashed line) cells without and with nitrate (39.5 mM) added to the RPMI 1640 medium. The data are averages of four experiments.

TABLE 2
Production of TNF (Measured by ELISA) by
Mouse P388D$_1$ Macrophages Transfected with a
Barley Nitrate Reductase Gene Fragment (NR)
and Treated with Nitrate (39.5 mM) for 24 Hours

Type of sample	TNF (ng/ml)
RPMI media control	
No nitrate	<0.05
With nitrate	0.05
P388D$_1$ cells (1 × 10^6)	
No nitrate	1.67
Cell extract	0.58
With nitrate	>3.20
P388D$_1$/NR cells (1 × 10^6)	
With nitrate	1.33

Furthermore, Figure 23 shows that the transfected P388D$_1$ cells, when exposed to nitrate, produced a cytokine that upregulated HIV in HeLa cells as measured by the production of chloramphenicol acetyltransferase activity by HeLa cells containing a HIV-LTR-CAT chromosomal insert.[625,626] The nitrate ion is an inhibitor of kinases and, therefore, may have produced an upregulation of NOS activity in control cells through this mechanism.[670] The cytokine produced by the P388D$_1$ cells was not nitrite because exposure to nitrite had no effect on the upregulation of HIV-LTR-CAT in the HeLa cells (see Figure 23). DALM production in animal cells to date has not been stimulated without acidification of nitrite.

III. SUMMARY

In nitrogen-metabolizing redox systems of bacteria, plants, and animals, there is a significant amount of similarity in structure and function. However, this similarity has been achieved in the wake of great diversity and in some instances accomplished by apparent parallel evolution. The survival of similar functions in the presence of great diversity attests to the ancient nature of the type-b cytochrome systems and their basic importance to life. It may also attest to a fundamental property that may allow the survival of function in a highly mutagenic environment (free radicals, chemical action, and radiation). There is no question that in the diverse forms of life in which type-b cytochrome systems are maintained, they occupy a pivotal position in metabolism and growth. The formation of composite redox systems from plant and mouse leukocyte components strongly suggests that in spite of the diversity, the conserved net structure and functions maintained in the nitrogen-metabolizing systems allow them to interact over great evolutionary distances.

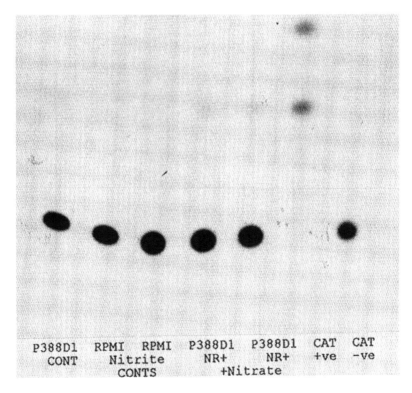

P388D1   RPMI   RPMI   P388D1   P388D1   CAT   CAT
CONT   Nitrite   NR+   NR+   +ve   -ve
CONTS   +Nitrate

FIGURE 23.   Response (24 hours) of HeLa HIV-LTR-CA. cells to supernatants from nitrate-reductase transfected P388D₁ mouse macrophages. This is an autoradiograph of the thin layer chromatography of the HeLa cell contents mixed with ¹⁴C-labeled chloramphenicol to detect solubility (migration) changes associated with chloramphenicol acetyltransferase activity. The results measure the degree of expression of the long-terminal-repeat sequence of HIV attached to a bacterial chloramphenicol acetyltransferase gene. +ve, = the positive control containing chloramphenicol acetyltransferase; –ve, = negative control; P388D₁ 1:20 (dilution), result in HeLa cells treated with P388D₁ supernatant not stimulated with nitrate; RPMI 1640 + NO₂⁻, medium controls with nitrite added (39.5 mM); P388D₁ NR⁺ + NO₂⁻, results from HeLa cells treated with supernatants from transfected P388D₁ cells treated with nitrate (39.5 mM).

Chapter 7

SUMMARY SCHEMATICS

CONTENTS

This chapter represents a collection of pictorial representations of the six previous chapters. It attempts to bring into sharp focus what is known and what is not known about the redox systems that have been previously explored in depth. The purpose is to give position and direction for present and future research, respectively.

$$2NADH \text{ OR } 2NADPH \xrightarrow{\text{Cytb}_5 \text{ Fp}} \begin{cases} 4Cyt\ b_5\ (Fe^{+3}) \rightarrow 4Cyt\ b_5\ (Fe^{+2}) \\ 4GHP\ (Fe^{+3}) \rightarrow 4GHP\ (Fe^{+2}) \end{cases}$$

$$4Cyt\ b_5\ (Fe^{+2}) + 4MetHb \longrightarrow Hb_{d4} + 4Cyt\ b_5\ (Fe^{+3})$$

$$4\ GHP\ (Fe^{+2}) + 4O_2 \text{ or } 8H_2O_2 + 4\ MetHb \longrightarrow Hb_4\ (O_2)_4 + 4\ GHP\ (Fe^{+3})$$

SCHEMATIC 1

The green hemoprotein of red blood cells as a superoxide methemoglobin transferase in red blood cells and its comparison to methemoglobin (cytochrome b_5) reductase. $Cytb_5Fp$ = methemoglobin (cytochrome b_5) reductase; $Cytb_5$ = soluble cytochrome b_5; $GHP(Fe^{+2} \text{ or } Fe^{+3})$ = green hemoprotein, reduced or oxidized, respectively; MetHb = methemoglobin; Hb_{d4} = deoxyhemoglobin; $Hb_4(O_2)_4$ = oxyhemoglobin.

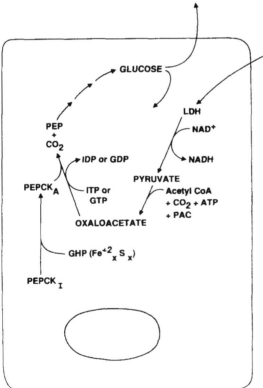

SCHEMATIC 2

The green hemoprotein, or phosphoenolpyruvate carboxykinase ferroactivator $(GHP(Fe^{+2}{}_xS_x))$, of hepatocytes. PEP = phosphoenolpyruvate; $PEPCK_A$ or $PEPCK_I$ = phosphoenolpyruvate carboxykinase, active or inactive forms, respectively; LDH, lactate dehydrogenase; ITP = inosine triphosphate; PAC = pyruvate carboxylase.

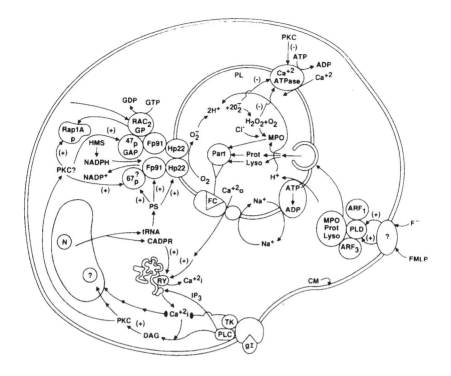

SCHEMATIC 3

NADPH oxidase in leukocytes. Part = particulate antigen; Lyso = lysozyme; Prot = protease; FC = Fc portion of immunoglobulin; MPO = myeloperoxidase; PKC = protein kinase C; PL = phagolysosome; ARF_1 and ARF_3 = degranulation co-factors; PLD = phospholipase D; F^- = fluoride ion, a metabolic stimulator; FMLP = N-formyl-L-methionyl-L-leucyl-L-phenylalanine, a chemotactic and oxidative-burst-stimulating peptide; Ca^{+2}_o or Ca^{+2}_i, extracellular and intracellular calcium ions, respectively; N = nucleus of cell; CM = cell membrane; RY = ryanodine receptor; IP_3 = inositol triphosphate; DAG = diacylglycerol; tRNA = transfer RNA; CADPR = cyclic adenosine diphosphate ribose; TK = tyrosine kinase; PLC = phospholipase C; gI = gamma-interferon; PS = protein synthesis; HMS = hexose monophosphate shunt; 47_p GAP = GTPase-activating component of NADPH oxidase; Rap1A = GTP-binding protein (G-protein) of NADPH oxidase; Rac_2 GP = a G-protein component of NADPH oxidase; 67_p = a cytoplasmic component of NADPH oxidase of unknown function; Fp91 = the flavoprotein component of the membrane-bound cytochrome b_{558} or b_{245}; Hp22 = the heme-binding component of cytochrome b_{558}; (+) or (–), positive or negative effects on the pathway, respectively.

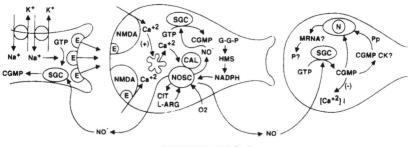

SCHEMATIC 4

Nitric oxide synthase in neurons. E = glutamate; SGC = soluble guanylate cyclase; CGMP = cyclic guanosine monophosphate; NMDA = N-methyl-D-aspartate; CAL = calmodulin; NOSC = constitutive neuronal nitric oxide synthase; CIT = citrulline; L-ARG = L-arginine; O_2 = oxygen; P = protein; Pp = phosphorylated protein; CGMP CK? = unidentified cGMP-dependent protein kinases; N = nucleus of cell; G-G-P = glucose to glucose-6-phosphate conversion; HMS = hexose monophosphate shunt; MRNA = messenger RNA; $[Ca^{+2}]^i$ = intracellular calcium ion concentration; NO˙ = nitric oxide.

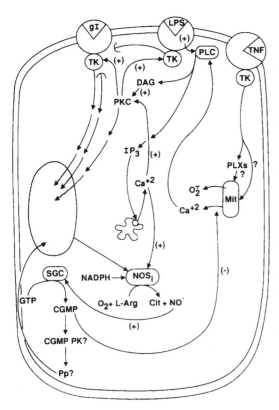

SCHEMATIC 5

Nitric oxide synthase of hepatocytes (hepNOS). TNF = tumor necrosis factor; LPS = lipopolysaccharide, or endotoxin; gI = gamma-interferon; PLC = phospholipase C; TK = tyrosine kinase; DAG = diacylglycerol; PKC = protein kinase C; IP_3 = inositol triphosphate; PLXs = variety of phospholipases (unidentified); NOS_i = hepNOS, or inducible hepatocyte nitric oxide synthase; L-ARG = L-arginine; Cit = citrulline; Mit = mitochondria; SGC = soluble guanylate cyclase; Pp? = phosphorylated protein, unidentified; CGMP PK? = unidentified cyclic guanylate monophosphate-dependent protein kinase; (+) or (−), positive or negative effects on the pathway, respectively.

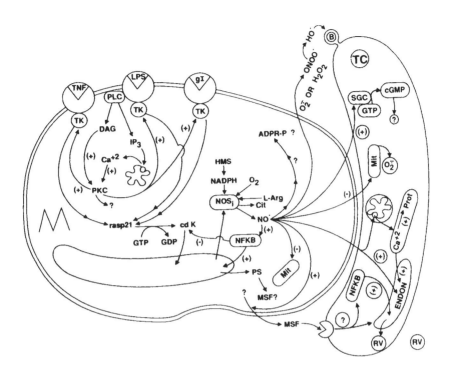

SCHEMATIC 6

Nitric oxide synthase in the macrophage being used to attack a target cell. M = macrophage; TC = target cell; TNF = tumor necrosis factor; LPS = lipopolysaccharide, or endotoxin; gI = gamma-interferon; TK = tyrosine kinase; PLC = phospholipase C; DAG = diacylglycerol; IP_3 = inositol triphosphate; rasp21 = a cytoplasmic G-protein; cd K = cyclin-dependent kinase involved in the cell cycle; NFKB = NF-kappa-B, a transcription factor; PKC = protein kinase C; NOS_i = macNOS, inducible macrophage nitric oxide synthase; HMS = hexose monophosphate shunt; L-Arg = L-arginine; Cit = citrulline; Mit = mitochondria; PS = protein synthesis; NO˙ = nitric oxide; ADPR-P = adenosine diphosphate ribosylated protein; MSF? = an unidentified macrophage stress factor; Prot = calcium ion-dependent protease; SGC = soluble guanylate cyclase; RV = retrovirus (latent, then expressed); ENDON = endonuclease, involved with apoptosis; B = bleb formation, associated with apoptosis.

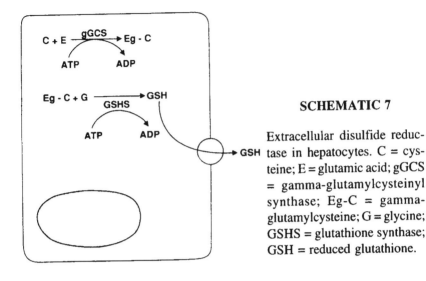

SCHEMATIC 7

Extracellular disulfide reductase in hepatocytes. C = cysteine; E = glutamic acid; gGCS = gamma-glutamylcysteinyl synthase; Eg-C = gamma-glutamylcysteine; G = glycine; GSHS = glutathione synthase; GSH = reduced glutathione.

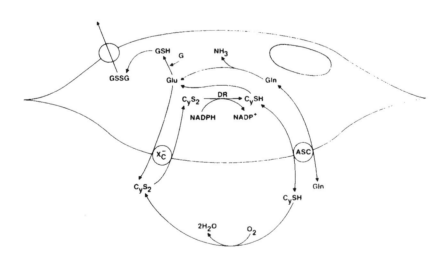

SCHEMATIC 8

Extracellular disulfide reductase in fibroblasts. DR = disulfide reductase; Gln = glutamine; Glu = glutamic acid; G = glycine; GSH = reduced glutathione; GSSG = oxidized glutathione; C_yS_2 = cystine; C_ySH = cysteine; X_C^- and ASC = amino acid transporters.

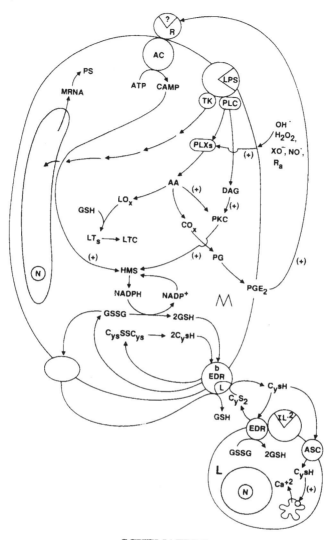

SCHEMATIC 9

Extracellular disulfide reductase in macrophages (M) and lymphocytes (L). R = receptor, unidentified; AC = adenylate cyclase; cAMP = cyclic adenosine monophosphate; mRNA = messenger RNA; PS = protein synthesis; LPS = lipopolysaccharide, or endotoxin; TK = tyrosine kinase; PLC = phospholipase C; PLXs = other phospholipases; XO^- = hypohalous acids; R_a = radiation (ionizing and nonionizing); DAG = diacylglycerol; PKC = protein kinase C; AA = arachidonic acid; LO_x = lipoxygenase; GSH = reduced glutathione; LT_s = leukotrienes; LTC = leukotriene C; CO_x = cyclooxygenase; PG = prostaglandins; PGE_2 = prostaglandin E_2; HMS = hexose monophosphate shunt; GSSG = oxidized glutathione; $C_{ys}SSC_{ys}$ or C_yS_2 = cystine; C_ysH = cysteine; b = cytochrome b component; EDR = extracellular disulfide reductase; L = unidentified ligand; IL^{-2} = interleukin 2; N = nucleus of cell; ASC = amino acid transporter.

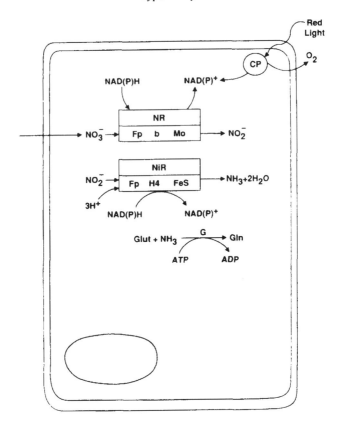

SCHEMATIC 10

Assimilatory nitrate and nitrite reductase. CP = chloroplast; NR = nitrate reductase; Fp = flavoprotein of nitrate reductase or nitrite reductase; b = cytochrome b component of nitrate reductase; Mo = molybdenum component of nitrate reductase; NiR = nitrite reductase; H4 = tetraheme component of nitrite reductase; FeS = nonheme-iron-sulfur component of nitrite reductase; G = glutamine synthetase.

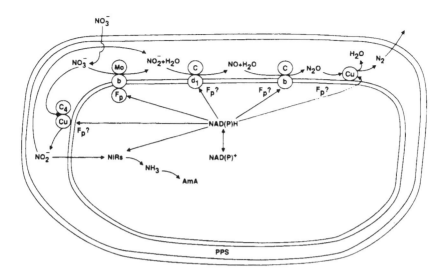

SCHEMATIC 11

Dissimilatory nitrate metabolism. NiRs = assimilatory nitrite reductase; AmA = amino acids; C_4 = tetracytochrome c component of periplasmic dissimilatory nitrate reductase; Cu = copper of periplasmic dissimilatory nitrate reductase or nitrous oxide reductase; C = cytochrome c components of dissimilatory nitrite reductase or nitric oxide reductase; b = cytochrome b components of nitrate or nitric oxide reductases; d_1 = d heme component of dissimilatory nitrite reductase; Mo = molybdenum-containing component of dissimilatory nitrate reductase; PPS = periplasmic space; Fp = flavoprotein component of nitrate reductase; Fp? = unidentified flavoprotein components of other nitrogen oxide reductases.

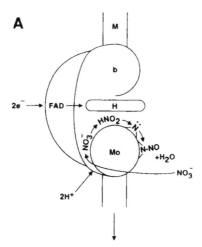

SCHEMATIC 12A

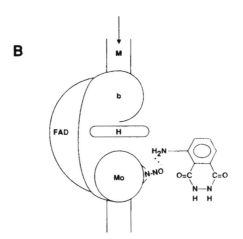

SCHEMATIC 12B

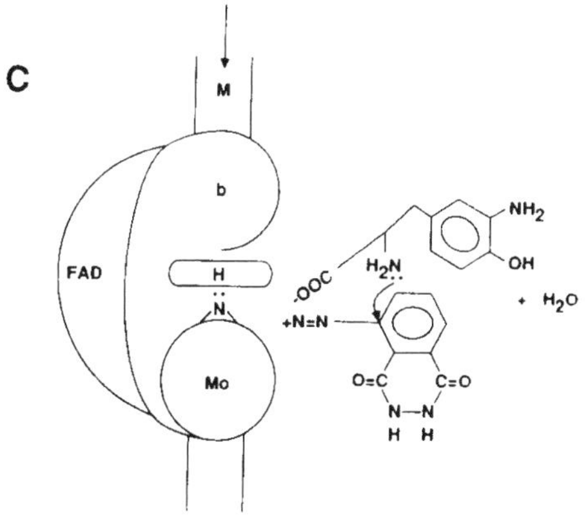

SCHEMATIC 12C

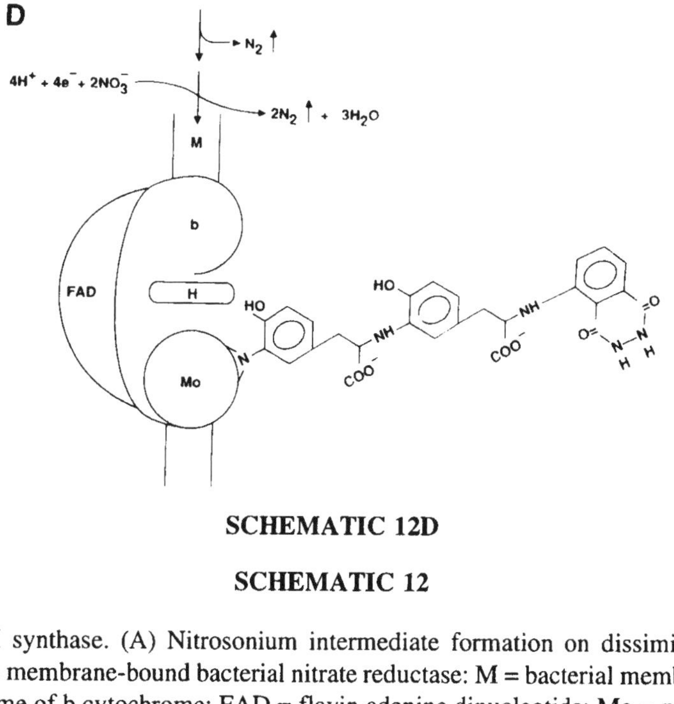

SCHEMATIC 12D

SCHEMATIC 12

DALM synthase. (A) Nitrosonium intermediate formation on dissimilatory aerobic membrane-bound bacterial nitrate reductase: M = bacterial membrane; H = heme of b cytochrome; FAD = flavin adenine dinucleotide; Mo = molybdenum-binding component; the arrows indicate the progression of the DALM-forming process. (B) The binding of luminol near the nitrosonium intermediate. (C) The diazotization of luminol and nucleophilic attack of the diazonium by the alpha amino group of 3-amino-L-tyrosine. (D) Repetitive diazotizations of the aromatic amines of 3-amino-L-tyrosines and displacement of the diazonium groups as nitrogen gas by the alpha amino groups of additional 3-amino-L-tyrosines to form the DALM polymer on the nitrate reductase.

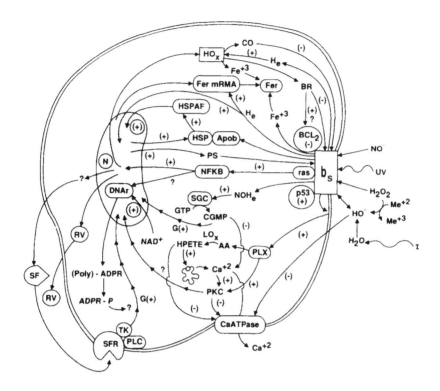

SCHEMATIC 13

The hypothetical type-b cytochrome sensor/switch. b_s = the type-b cytochrome sensor; UV = ultraviolet light; Me^{+2} or Me^{+3} = reduced and oxidized transition, or heavy-metal ions; I = ionizing radiation; BR = bilirubin; HO_x = heme oxygenase; CO = carbon monoxide; H_e = heme; PS = protein synthesis; Fer = ferritin; PLX = unidentified phospholipase; HSPAF = heat shock protein activating factor; HSP = heat shock protein; Apob = cytochrome b without a heme; NOH_e = nitric oxide bound to free heme; SGC = soluble guanylate cyclase; HPETE = hydroperoxyeicosatetranoic acids; PKC = protein kinase C; LO_x = lipoxygenase; G(+) = positive growth effects; TK = tyrosine kinase; ADPR-P = adenosine diphosphate ribosylated protein; (Poly)-ADPR = poly-adenosine diphosphate ribose; AA = arachidonic acid; DNAr = DNA repair; BCL_2 = the gene product BCL_2 has a negative effect on the oxidative activity of the sensor; p52(+) = the gene product p53 has a positive effect on the sensor; SF = stress factor; SFR = stress factor receptor; RV = retrovirus such as HIV; N = cell nucleus.

Chapter 8

FUTURE RESEARCH AND APPLICATIONS

CONTENTS

I. FUTURE RESEARCH AND APPLICATIONS QUESTIONS

This final chapter is a summary of the questions yet to be answered in the field of cytochrome b structure and functions, and the applications toward which the research is leading us. The principal question that this monograph addressed but did not completely answer is this: Does a nonspecific cell sensor exist, and if so, what and where is it? The apparent answer, based upon the data examined, is yes, it does exist. It is a singular entity or family of closely related entities that translates many diverse nonspecific stimuli (temperature change, pH change, radiation including visible light, UV, and ionizing radiation) and specific stimuli (receptor interactions, antigenic stimulation, cytokine and mitogen interactions) into a reductive/oxidative metabolic response. The sensor(s) is preferentially in the cytoplasmic membrane in order that it may respond to external stimuli before lethal or irreversible internal damage is experienced. The sensor must be very sensitive to external stimuli in order to respond early enough for the cell to generate a protective response. Without a successful preventative response, there would be no evolutionary drive to offset lethal stimuli. Therefore, the sensor must switch on responses that are not only adequate for responding to previously experienced challenges, but it must also generate excessive responses that anticipate future, more robust, challenges. Without the capacity for the latter, there would be no survivors to further evolve. Yet this sensor must be tightly linked to cell growth initiation and other internal metabolic events and also must be able to provide for communication with surrounding cells.

In this book, three candidate pathway types that may contain or be linked to the sensor were examined: (1) the superoxide/hydrogen peroxide/hydroxyl radical-producing pathway of NADPH oxidase; (2) the nitrogen oxide producing and metabolizing pathways of NO synthase and nitrogen oxide reductases; (3) the extracellular/intracellular thiol-balancing pathway of extracellular disulfide reductase. Three types of redox proteins and/or polypeptides are central to all three of these metabolic pathway types: (1) flavoprotein; (2) type-b cytochromes (and substitutes); and (3) protein and/or polypeptide thiols (including disulfide reductases, isomerases, and transhydrogenases).

In NADPH oxidase and NO synthases, these three components are so highly integrated that two are fused into the same protein, forming flavocytochromes. NADPH oxidase has both membrane and cytoplasmic components, making it a good candidate for the sensor. The membrane component is a type-b cytochrome of sufficient instability to be a sensor candidate. Also, as noted in Chapter 3, the mRNA for this cytochrome (b_{558}) is produced in nonleukocytic cells without active (or complete) NADPH oxidase.

NO synthase would be a good candidate for the sensor because it is widespread in cell types (neurons, leukocytes, and other cells), but it is in the active form only as a cytoplasmic enzyme, is activated after calcium release (constitutive) or after a long induction period (>4 hours) necessary for protein synthesis. The constitutive form, unlike the inducible form, can be readily turned off by calcium sequestration. A major question remains as to how the inducible nitric oxide synthase is turned off (decreased synthesis, increased degradation, and/or self-destruction of the sensitive cytochrome b component?).

The redox sensor activation must precede the release or mobilization of calcium. Additionally, the fact that redox events and reducing equivalent depletion affect calcium mobilization in cells suggests that a redox event precedes the calcium mobilization and that such an event if tied to a slow fluorescence mechanism would be the most likely target for weak electromagnetic field (or radiation) effects. Also, programmed cell death or apoptosis is involved with a hydroxyl-radical membrane event.

The intense interest in weak electromagnetic fields and radiofrequency radiation effects has centered on calcium mobilization. However, the required binding energies and transport energy requirements preclude weak field effects. Redox sensors, based on quantum electrodynamics, are potentially the ultimate biosensors required for sensitivity to such weak influences because singlet/triplet intersystem crossing (spin flipping) between states of the same energy level are "lossless", requiring neither a gain nor a loss of energy. The fact that such processes are operational in photosynthesis and in electron transfer in mitochondria strongly suggests that developing an ideal "lossless" sensor is a goal toward which evolution is being driven.

Therefore, the sensor should be like the type-b cytochrome of NADPH oxidase or of the membrane-bound nitrate reductase of bacteria. The role of BCL-2 in inhibition of the hydroxyl-free-radical-apoptotic event needs to be clarified as well as the p53 enhancement of apoptosis (to determine if it is a redox event or not) because they may be involved with free radicals generated by such enzymes. However, the sensor should generate rapid communication (intracellular and extracellular) like NO synthase through NO. The only currently known receptor for NO is soluble guanylate cyclase, or specifically, heme bound to the soluble guanylate cyclase. However, as noted in Chapter 4, cells without soluble guanylate cyclase also respond to NO. Therefore, the pathways that control ADP ribosylation of proteins and their components that interact with NO and yield the effects of NO need further investigation. The search for heretofore undiscovered NO responsive systems also needs to be pursued.

Along this line, the activation of retroviruses such as HIV by NO needs further examination. This requirement is especially urgent in light of the paradox created by the observations that other viruses are inactivated and other infectious agents are killed or inhibited by NO.

The extracellular disulfide reductase is the third candidate for being the sensor. It seems critical to at least several ligand/antiligand (receptor) interactions, activation of latent Abelson leukemia virus in RAW 264.7 cells, activation of lymphocyte proliferation (the first steps in the specific immune response), and cell survival and proliferation after mitogenic or nonspecific stressor stimulation. Candidates have been found for extracellular disulfide reductase: (1) thioredoxin; (2) follicle stimulating hormone; (3) luteinizing hormone; (4) undefined cytokines; (5) the cytochrome b component of NADPH oxidase; and (6) intramembrane thiol (amino acid) transporters. However, the issue remains unsettled and the source of reducing equivalents undefined. Is it a thiol transhydrogenase or isomerase or an intracellular flavoprotein or flavocytochrome that transfers electrons to a membrane component that, in turn, supplies them to extracellular substrates? Such a system could be identical with the membrane-bound cellular NAD(P)H diaphorase of neurons and leukocytes. The soluble NO synthase could form the intracellular cytoplasmic reductase that binds to the membrane component in the assembly of the extracellular disulfide reductase, or the NO synthase could be a proteolytic soluble product of the precursor membrane-bound extracellular disulfide reductase. This question needs to be settled.

The nitrogen oxide reductases of bacteria provide a model for such transmembrane directional (extracellularly directed) responding sensors. Their importance as sensors is confirmed by their roles in controlling bacterial anoxic respiration. In higher plants, the cytoplasmic soluble nitrogen oxide reductase (nitrate and nitrite) show hysteretic behavior that suggests that they too are sensors of redox status. Their importance in this role is demonstrated by their linkages to photosynthesis and mitochondrial respiration.

Preliminary data on the integration of a plant nitrate reductase component with NO synthase of mouse macrophages (see Chapter 6) suggest that the NO synthase/cytochrome P450 could be linked to a redox sensor. The effects of this system on HIV expression do suggest that nitrogen oxide metabolism leads to intercellular (cytokine?) communication that enhances HIV and other latent retrovirus upregulation. Further work is needed to understand direct interactions, if any, of NADPH oxidase, NO synthase, and extracellular disulfide reductase components in cells. These investigations will lead us closer to defining a cellular environmental sensor and switch as a collective effect of redox systems or a specific component of the cell membrane.

II. FUTURE APPLICATIONS

Even with so many open research questions to still be addressed, applications of the knowledge rapidly accumulating in this area are becoming apparent. In the forefront of potential applications is the development of strategies for prevention of and therapy for acquired immunodeficiency syndrome (AIDS). The importance of new strategies and paradigms in approaching AIDS prevention and treatment was emphasized by Albert B. Sabin in his last letter to *Nature* in 1993, just prior to his death.[671]

He emphasized the "novel" transfer of HIV by cell-to-cell contact without free extracellular virus during sexual or blood-borne transmission. He stated that antibody and cell-mediated immunity were inadequate in preventing the accumulation in lymphoid and other tissues of a reservoir of HIV-infected cells (many latently infected with viral cDNA). In Chapter 5, results were discussed of pathological examinations that provided evidence for widespread latent infection by HIV of lymphoid tissues in humans prior to evidence of the virus in plasma or peripheral blood mononuclear cells or symptoms of disease. Sabin raised doubt about whether modified viral vaccines would be any better at protecting individuals from AIDS than the natural virus (Sabin's law).

As noted in Chapters 4 and 5, HIV and its components seem to trigger pathophysiologic changes in both immune and nonimmune tissues that are directly linked to redox metabolism. NO synthase activation favors HIV reactivation (expression), and extracellular disulfide reductase activation and accompanying release of undefined cytokine(s) (some of which normally favor cell survival under environmentally stressful conditions) favor viral expression, replication, and infection. One might consider using tumor necrosis factor (TNF) or lipopolysaccharide (LPS) and gamma-interferon to stimulate the immune system to inactivate HIV. However, based on the results discussed in this book, these immunostimulants are expected to accelerate viral expression and AIDS pathology.[672] Vaccines as specific immunostimulants may have the same effects. On the other hand, anti-inflammatory agents and immunosuppressive drugs that may inhibit these metabolic processes would likely lead to the same net immunosuppressive effects of the viral infection (although they might relieve the nonimmunologic symptoms such as neurodegeneration).

This dilemma leads applied researchers and clinicians to two general strategies for prevention and therapy: (1) prevention of entry of virus at the body surfaces (physical barriers such as condoms and screened blood products and immunologic barriers such as mucosal vaccines), and (2) prevention of expression of virus in latently infected cells. Based on Sabin's observations and suggestions, the only feasible vaccine would be an IgA-inducing one that prevents entry through mucosal surfaces. In individuals already infected, the use of ribozymes directed toward the viral RNA components in the cell would support the second strategy. Where viral mRNA is being processed, the ribozymes would have access to them for the purpose of cutting them into fragments that could be neither packaged in virions nor translated into proteins.[673] The viral-linked ribozyme would be coexpressed when the latent virus is activated (expressed as mRNA). This "tit-for-tat" approach would arrest expression and further infection of other cells by the virus. Hopefully, it would also prevent symptomology associated with cell death resulting from viral replication and redox metabolic activation. However, in light of the widespread latent infection of lymphoid and other tissues before the appearance of AIDS, this approach yields a stalemate rather than a cure, at best. If all these latently infected cells were killed, which is the usual approach of cellular immunity to eliminate viruses from the body, then the net effect would be AIDS.

To "cure" HIV-infected cells, the viral cDNA in the cell genome needs to be removed. To accomplish this, an endonuclease would be required to attack the segment of viral cDNA and a ligase would be required to repair the break by reannealing the DNA after excision and deletion in both strands of DNA. The best theoretical approach to this problem would be to initiate the excision and nuclease activity at the time of transcription of the viral genes. Nitric oxide may still be the agent of choice for this process, if whatever process that protects the virus from naturally induced NO production (cytokine or thiols?) in the cell can be overcome without killing the cell with NO toxicity. If NO's abilities to stop cell growth and initiate viral expression could be preserved while still maintaining its nucleic acid-deaminating and hydroxyl-radical-forming properties, then it could be directed toward viral DNA destruction.

The first step toward gaining external control over NO synthase activity in specific cell types (lymphocytes and macrophages) was made with the transfection of barley nitrate reductase (fragment) into mouse macrophages (see Chapter 6). The expression of these fragments allowed for control over NO production by feeding exogenous nitrate (a usually innocuous substrate in adult humans). The cells demonstrated cytostasis or cytocidal effects, depending upon the level of upregulation of activity with nitrate (dose response). The nonfunctionality of the fragments in cells lacking an active NO synthase but transfected with the gene fragments indicated that specificity could be accomplished by the properties of the host cell. What has not been achieved yet is specific control within a permissive host cell, other than by nitrate levels. For HIV, this specificity could be achieved by fusing the HIV-LTR-TAT genes with the nitrate reductase gene fragment and placing it as cDNA in the host cell genome. Therefore, when the HIV transcription is induced, the fused DNA-binding protein TAT and the nitrate reductase will bind to the transcribing HIV cDNA. When nitrate is also present, nitric oxide and nitrite will be formed at the transcriptional unit while other host cell genes are quiet. Also, the nitrate reductase expression would follow the "tit-for-tat" dosimetric approach discussed for ribozyme therapy.

This general approach of multiple control points for nitric oxide/nitrite production could be applied to a number of "genetic surgery" problems including removal of defective oncogenes or other latent viruses. The approach could also convey resistance (to NO killing or inactivation mechanisms) or sensitivity (self-intoxication) to various chemicals, infectious agents, or physical stressors that induce certain genes. Insights into and exploitation of the redox basis of programmed cell death, or apoptosis, and the involvement in these processes of p53 and BCL-2 would act to facilitate amplification or inhibition of these sensitivities. Vector-borne recombinant nitrate/nitrite reductases could also be used to treat cardiovascular disease (bringing about focused vasodilation) in conjunction with a much lower dose of nitrate than that of the exogenous nitrocompounds usually administered.

Environmentally, the formation of nitrites and nitric oxide induced by recombinant DNA constructs responsive to a variety of physical and chemical stressors would form biosensors for all manner of environmental agents. Such recombinant nitrogen-oxide-reductase biosensors could be made more sensitive and rapid by coupling diazoluminomelanin formation to the gene expression. Detection of the appearance of slow fluorescence/luminescence in these specifically engineered bacteria or eukaryotic cells would be a very sensitive assay system for the specific expression response of the cells.

The microwave-absorbing and enhanced heating of such engineered cells *in situ, in vivo,* or of their products such as antibody or proteins linked to the DALM by radiofrequency radiation sources would add a new dimension to the focusing of penetrating microwave or radiowave radiation therapies. The therapeutic uses would include hyperthermic treatment of cancer, radiofrequency radiation surgery and killing of infectious agents.

Finally, again in the environmental area, the genetic linking of various redox enzymes or cytochromes to produce hybrid composite systems could lead to new metabolism for the breakdown of nitroaromatics, nitroaliphatics, halocarbons, and other environmental toxicants and environmental-quality-degrading compounds.

REFERENCES

1. **Kauffmann, W. J., III,** *Universe,* 4th Ed., W. H. Freeman, New York, 1994, chaps. 8 and 12.

2. **Leovy, C. B.,** The atmosphere of Mars, in *The Planets,* Murray, B., Ed., W. H. Freeman, New York, 1983, chap. 4.

3. **Arvidson, R. E., Binder, A. B., and Jones, K. L.,** The surface of Mars, in *The Planets,* Murray, B., Ed., W. H. Freeman, New York, 1983, chap. 5.

4. **Dose, K., and Zaki, L.,** Hamoproteinoide mit peroxidatischer und katalatischer aktivitat, *Zeitschrift fuer Naturforschung,* 26b, 144, 1971.

5. **Welinder, K. G.,** Plant peroxidases: their primary, secondary and tertiary structures, and relation to cytochrome c peroxidase, *European Journal of Biochemistry,* 151, 497, 1985.

6. **Kimura, S., and Yamazaki, I.,** Heme-linked ionization and chloride binding in intestinal peroxidase and lactoperoxidase, *Archives of Biochemistry and Biophysics* 189, 14, 1978.

7. **Lazar, J. G., and Ross, J.,** Changes in mean concentration, phase shifts, and dissipation in a forced oscillatory reaction, *Science,* 247, 189, 1990.

8. **Morris, D. R., and Hager, L. P.,** Chloroperoxidase: isolation and properties of the crystalline glycoprotein, *Journal of Biological Chemistry,* 241, 1763, 1966.

9. **Degn, H.,** Compound III kinetics and chemiluminescence in oscillatory oxidation reactions catalyzed by horseradish peroxidase, *Biochimica et Biophysica Acta,* 180, 271, 1969.

10. **Sitter, A. J., Reczek, C. M., and Terner, J.,** Comparison of the heme structures of horseradish peroxidase compounds X and II by resonance Raman spectroscopy, *Journal of Biological Chemistry,* 261, 8638, 1986.

11. **Sono, M., Eble, K. S., Dawson, J. H., and Hager, L. P.,** Preparation and properties of ferrous chloroperoxidase complexes with dioxygen, nitric oxide, and an alkyl isocyanide: spectroscopic dissimilarities between the oxygenated forms of chloroperoxidase and cytochrome P-450, *Journal of Biological Chemistry,* 260, 15530, 1985.

12. **Shannon, L. M., Kay, E., and Lew, J. Y.,** Peroxidase isozymes from horseradish roots, *Journal of Biological Chemistry,* 241, 2166, 1966.

13. **Nakajima, R., and Yamazaki, I.,** The mechanism of indole-3-acetic acid oxidation by horseradish peroxidase, *Journal of Biological Chemistry,* 254, 872, 1979.

14. **Hollenberg, P. F., Rand-Meir, T., and Hager, L. P.,** The reaction of chlorite with horseradish peroxidase and chloroperoxidase, *Journal of Biological Chemistry,* 249, 5816, 1974.

15. **Haun, M., Duran, N., Augusto, O., and Cilento, G.,** Model studies of the alpha-peroxidase system: formation of an electronically excited product, *Archives of Biochemistry and Biophysics,* 200, 245, 1980.

16. **Yokota, K.-N., and Yamazaki, I.,** Analysis and computer simulation of aerobic oxidation of reduced nicotinamide adenine dinucleotide catalyzed by horseradish peroxidase, *Biochemistry,* 16, 1913, 1977.

17. **Takayama, K., and Nakano, M.,** Mechanism of thyroxine-mediated oxidation of reduced nicotinamide adenine dinucleotide in peroxidase-H_2O_2 system, *Biochemistry,* 16, 1921, 1977.

18. **Yamazaki, I., and Piette, L. H.,** The mechanism of aerobic oxidase reaction catalyzed by peroxidase, *Biochimica et Biophysica Acta,* 77, 47, 1963.

19. **Araiso, T., and Dunford, H. B.,** Horseradish peroxidase. Complex formation with anions and hydrocyanic acid, *Journal of Biological Chemistry,* 256, 10099, 1981.

20. **Dawson, J. H.,** Probing structure-function relations in heme-containing oxygenases and peroxidases, *Science,* 240, 443, 1988.

21. **Wittenberg, J. B., Noble, R. W., Wittenberg, B. A., Antonini, E., Brunori, M., and Wyman, J.,** Studies on the equilibria and kinetics of the reactions of peroxidase with ligands. II. The reaction of ferroperoxidase with oxygen, *Journal of Biological Chemistry,* 242, 626, 1967.

22. **Nakajima, R., and Yamazaki, I.,** The conversion of horseradish peroxidase C to a verdohemoprotein by a hydroperoxide derived enzymatically from indole-3-acetic acid and by *m*-nitroperoxybenzoic acid, *Journal of Biological Chemistry,* 255, 2067, 1980.

23. **Halliwell, B., and Gutteridge, J. M. C.,** *Free Radicals in Biology and Medicine,* Clarendon Press, Oxford, 1985, chaps. 1–8.

24. **Koppenol, W. H.,** Generation and theromodynamic properties of oxyradicals, in *Membrane Lipid Oxidation,* Vol. I, Vigo-Pelfrey, C., Ed., CRC Press, Boca Raton, FL, 1990, chap. 1.

25. **Palanioppan, V., and Terner, J.,** Resonance Raman spectroscopy of horseradish peroxidase derivatives and intermediates with excitation in the near ultraviolet, *Journal of Biological Chemistry,* 264, 16046, 1989.

26. **Stelmakh, G. F., and Tsvirko, M. P.,** Delayed fluorescence from the upper excited electronic states of metalloporphyrins, *Optical Spectroscopy* (USSR), 49, 278, 1980.

27. **Feltelson, J., and Mauzerall, D.,** Reactions of triplet states of a porphyrin measured by delayed fluorescence, *Journal of Physical Chemistry,* 86, 1623, 1982.

28. **Gething, M.-J., and Sambrook, J.,** Protein folding in the cell, *Nature,* 355, 33, 1992.

29. **Rivera, M. C., and Lake, J. A.,** Evidence that eukaryotes and eocyte prokaryotes are immediate relatives, *Science,* 257, 98, 1981.

30. **Woese, C. R.,** Archebacteria, *Scientific American,* 244, 98, 1981.

31. **Flam, F.,** Finding RNA makes proteins gives "RNA world" a big boost, *Science,* 256, 1396, 1992.

32. **Pace, N. R.,** New horizons for RNA catalysis, *Science,* 256, 1402, 1992.

33. **Reichard, P.,** From RNA to DNA, why so many ribonucleotide reductases? *Science,* 260, 1773, 1993.

34. **Metzler, D. E.,** *Biochemistry: The Chemical Reactions of Living Cells,* Academic Press, New York, 1977, chap. 10.

35. **Metzler, D. E.,** *Biochemistry: The Chemical Reactions of Living Cells,* Academic Press, New York, 1977, chap. 13.

36. **Randall, L. L.,** Peptide binding by chaperone secB: implications for recognition of nonnative structure, *Science,* 257, 241, 1992.

37. **Beckmann, R. P., Mizzen, L. A., and Welch, W. J.,** Interaction of hsp70 with newly synthesized proteins: implications for protein folding and assembly, *Science,* 248, 850, 1990.

38. **Neupert, W., and Schatz, G.,** How proteins are transported into mitochondria, *Trends in Biological Science,* Jan, 1, 1981.

39. **Hartl, F.-U., and Neupert, W.,** Protein sorting to mitochondria: evolutionary conservations of folding and assembly, *Science,* 247, 930, 1990.

40. **Mayo, S. L., Ellis, W. R., Jr., Crutchley, R. J., and Gray, H. B.,** Long-range electron transfer in heme proteins, *Science,* 233, 948, 1986.

41. **Kuo, C.- F., McRee, D. E., Fisher, C. L., O'Handley, S. F., Cunningham, R. P., and Tainer, J. A.,** Atomic structure of the DNA repair [4Fe-4S] enzyme endonuclease III, *Science,* 258, 434, 1992.

42. **De Mello, M. P., De Toledo, S. M., Haun, M., Cilento, G., and Duran, N.,** Excited indole-3-aldehyde from the peroxidase-catalyzed aerobic oxidation of indole-3-acetic acid. Reaction with and energy transfer to transfer ribonucleic acid, *Biochemistry,* 19, 5270, 1980.

43. **Morishima, I., Kurono, M., and Shiro, Y.,** Presence of endogenous calcium ion in horseradish peroxidase. Elucidation of metal-binding site by substitution of divalent and lanthanide ions for calcium and use of metal-induced NMR (^{1}H and ^{113}Cd) resonances, *Journal of Biological Chemistry,* 261, 9391, 1986.

44. **Shiro, Y., Kurono, M., and Morishima, I.,** Presence of endogenous calcium ion and its functional and structural regulation in horseradish peroxidase, *Journal of Biological Chemistry,* 261, 9382, 1986.

45. **Sticher, L., Penel, C., and Greppin, H.,** Calcium requirement for the secretion of peroxidases by plant cell suspensions, *Journal of Cell Science,* 48, 345, 1981.

46. **Jellinck, P. H., and McNabb, T.,** Metabolism of estradiol by uterine peroxidase: nature of the water soluble products, *Steroids,* 29, 525, 1977.

47. **Lyttle, C. R., and Jellinck, P. H.,** Metabolism of [4-^{14}C] oestradiol by oestrogen-induced uterine peroxidase, *Biochemical Journal,* 127, 481, 1972.

48. **Norymberski, J. K.,** Enzymic and biomimetic oxidative coupling of oestrogens, *FEBS Letters,* 86, 163, 1978.

49. **Lin, J. J., Daniels-McQueen, S., Patino, M. M., Gaffield, L., Walden, W. E., and Thach, R. E.,** Derepression of ferritin messenger RNA translation by hemin *in vitro,* *Science,* 247, 74, 1990.

50. **Cosma, G., Fulton, H., De Feo, T., and Gordon, T.,** Rat lung metallothionein and heme oxygenase gene expression following ozone and zinc oxide exposure, *Toxicology and Applied Pharmacology,* 117, 75, 1992.

51. **Kikuchi, G., and Yoshida, T.,** Heme degradation by the microsomal heme oxygenase system, *Trends in Biological Science,* December, 323, 1980.

52. **Tyrrell, R. M., Keyse, S. M., Applegate, L. A., and Loutier, D.,** Inducible cellular defense against oxidative stress in cultured human cells, in *Oxidative Damage and Repair. Chemical, Biological and Medical Aspects,* Davies, K. J. A., Ed., Pergamon Press, Oxford, 1991, 19.

53. **Hochstein, P., Atallah, A. S., and Ernster, L.,** DT diaphorase and toxicity of quinones: status and perspectives, in *Cellular Antioxidant Defense Mechanisms,* Vol. II, Chow, C. K., Ed., CRC Press, Boca Raton, FL, 1988, chap. 21.

54. **Bentle, L. A., and Lardy, H. A.,** *p*-Enolpyruvate carboxykinase ferroactivator. Purification and some properties, *Journal of Biological Chemistry,* 252, 1431, 1977.

55. **Doroshow, J. H.,** Role of oxygen radical metabolism in the cardiac toxicity of anticancer quinones, in *Membrane Lipid Oxidation,* Vol. I, Vigo-Pelfrey, C., Ed., CRC Press, Boca Raton, FL, 1990, chap. 14.

56. **Rich, P. R., Moody, A. J., and Mitchell, R.,** The charge translocating reactions of the bc complexes and cytochrome oxidase, in *Charge and Field Effects in Biosystems,* Vol. 2, Allen, M. J., Cleary, S. F., and Hawkridge, F. M., Eds., Plenum Press, New York, 1989, 7.

57. **Parkos, C. A., Allen, R. A., Cochrane, C. G., and Jesaitis, A. J.,** The quarternary structure of the plasma membrane b-type cytochrome of human granulocytes, *Biochimica et Biophysica Acta,* 932, 71, 1988.

58. **Esposti, M. D.,** Prediction and comparison of the haem-binding sites in membrane haemoproteins, *Biochimica et Biophysica Acta,* 977, 249, 1989.

59. **Babior, B. M.,** The components of the respiratory burst oxidase of human neutrophils, in *Oxidative Damage and Repair: Chemical, Biological and Medical Aspects,* Davies, K. J. A., Ed., Pergamon Press, Oxford, 1991, 647.

60. **Madison, D. V.,** Pass the nitric oxide, Proceedings of the *National Academy of Sciences of the USA,* 90, 4329, 1993.

61. **Thauer, R. K., Jungermann, K., and Decker, K.,** Energy conservation in chemotrophic anaerobic bacteria, *Bacteriological Reviews,* 41, 100, 1977.

62. **De Filippi, L. J., Ballow, D. P., and Hultquist, D. E.,** Reaction of bovine erythrocyte green hemoprotein with oxygen and hydrogen peroxide, *Journal of Biological Chemistry,* 254, 6917, 1979.

63. **Kiel, J. L., McQueen, C., and Erwin, D. N.,** Green hemoprotein of erythrocytes: methemoglobin superoxide transferase, *Physiological Chemistry and Physics and Medical NMR,* 20, 123, 1988.

64. **De Filippi, L. J., and Hultquist, D. E.,** The green hemoprotein of bovine erythrocytes. I. purification and characterization, *Journal of Biological Chemistry,* 253, 2946, 1978.

65. **Galeris, D. , Buffinton, G., Hochstein, P., and Cadenas, E.,** Role of ferry myoglobin in lipid peroxidation and its reduction to met- or oxymyoglobin by glutathione, quinones, quinone thioether derivatives, and ascorbate, in *Membrane Lipid Oxidation,* Vol. I, Vigo-Pelfrey, C., Ed., CRC Press, Boca Raton, FL, 1990, chap. 12.

66. **White, A., Handler, P., and Smith, E. L.,** *Principles of Biochemistry,* 4th Ed., McGraw-Hill, New York, 1968, chap. 8.

67. **Shikama, K., and Sugawara, Y.,** Autoxidation of native oxymyoglobin. Kinetic analysis of the pH profile, *European Journal of Biochemistry,* 91, 407, 1978.

68. **Pommier, J., and Cahnmann, H. J.,** Interaction of lactoperoxidase with thiols and diiodotyrosine, *Journal of Biological Chemistry,* 254, 3006, 1979.

69. **Ikeda-Saito, M.,** Spectroscopic, ligand binding, and enzymatic properties of the spleen green hemeprotein. A comparison with myeloperoxidase, *Journal of Biological Chemistry,* 260, 11688, 1985.

70. **Sharma, V. S., and Traylor, T. G.,** Why NO? *Biochemistry,* 31, 2847, 1992.

71. **Heath, R. L., and Packer, L.,** Photoperoxidation in isolated chloroplasts. I. Kinetics and stoichiometry of fatty acid peroxidation, *Archives of Biochemistry and Biophysics,* 125, 189, 1968.

72. **Gallop, P. M.,** The biological significance of quinonoid compounds in, on, and out of proteins, *Chemtracts-Biochemistry and Molecular Biology,* 1, 357, 1990.

73. **Loschen, G., Azzi, A., Richter, C., and Flohe, L.,** Superoxide radicals as precursors of mitochondrial hydrogen peroxide, *FEBS Letters,* 42, 68, 1974.

74. **Pan, S.-S., and Nason, A.,** Purification and characterization of homogeneous assimilatory reduced nicotinamide adenine dinucleotide phosphate-nitrate reductase from *Neurospora crassa, Biochimica et Biophysica Acta,* 52 , 297, 1978.

75. **Kiel, J. L., and Erwin, D. N.,** Microwave radiation effects on the thermally driven oxidase of erythrocytes, *International Journal of Hyperthermia,* 2, 201, 1986.

76. **Heber, U., Kirk, M. R., and Boardman, N. K.,** Photoreactions of cytochrome b-559 and cyclic electron flow in photosystem II of intact chloroplasts, *Biochimica et Biophysica Acta,* 546, 292, 1979.

77. **Itoh, S., and Iwaki, M.,** Delayed fluorescence in photosystem I enhanced by phylloquinone (vitamin K-1) extraction with ether, *Biochimica et Biophysica Acta,* 934, 32, 1988.

78. **Cooper, A. J. L.,** L-Citrulline production from L-arginine by macrophage nitric oxide synthase: the ureido oxygen derives from dioxygen, *Chemtracts—Biochemistry and Molecular Biology,* 1, 404, 1990.

79. **Bredt, D. S., and Snyder, S. H.,** Isolation of nitric oxide synthetase, a calmodulin-requiring enzyme, *Proceedings of the National Academy of Sciences of the USA,* 87, 682, 1990.

80. **Gabig, T. G., and Babior, B. M.,** The O_2^--forming oxidase responsible for the respiratory burst in human neutrophils. Properties of the solubilized enzyme, *Journal of Biological Chemistry,* 254, 9070, 1979.

81. **Morel, F., and Vignais, P. V.,** Purification of cytochrome b_{558} from bovine polymorphonuclear neutrophils, *Biochemical and Biophysical Research Communications,* 149, 46, 1987.

82. **Davis, J. C., and Averill, B. A.,** Isolation from bovine spleen of a green heme protein with properties of myeloperoxidase, *Journal of Biological Chemistry,* 256, 5992, 1981.
83. **Black, M. T., Widger, W. R., and Cramer, W. A.,** Large-scale purification of active cytochrome b₆/f complex from spinach chloroplasts, *Archives of Biochemistry and Biophysics,* 252, 655, 1987.
84. **Junge, W., Hang, Y.-Q., Theg, S., Forster, V., and Polle, A.,** Localized protons in photosynthesis of green plants?, in *Information and Energy Transduction in Biological Membranes,* Alan R. Liss, New York, 1984, 139.
85. **Heinz, E.,** Some basic problems of proton-ATPase: chairman's introduction, in *Information and Energy Transduction in Biological Membranes,* Alan R. Liss, New York, 1984, 167.
86. **Lauger, P.,** Mechanistic properties of ion channels and pumps, in *Information and Energy Transduction in Biological Membranes,* Alan R. Liss, New York, 1984, 133.
87. **West, I. C., Mitchell, P., and Rich, P. R.,** Electron conduction between b cytochromes of the mitochondrial respiratory chain in the presence of antimycin plus myxothiazol, *Biochimica et Biophysica Acta,* 933, 35, 1988.
88. **Rich, P. R., Jeal, A. E., Madgwick, S. A., and Moody, A. J.,** Inhibitor effects on redox-linked protonations of the b haeme of the mitochondrial bc₁ complex, *Biochimica et Biophysica Acta,* 1018, 29, 1990.
89. **Metzler, D. E.,** *Biochemistry: The Chemical Reactions of Living Cells,* Academic Press, New York, 1977, chap. 11.
90. **Metzler, D. E.,** *Biochemistry: The Chemical Reactions of Living Cells,* Academic Press, New York, 1977, chap. 7.
91. **Yokota, K.-N., and Yamazaki, I.,** The activity of the horseradish peroxidase compound III, *Biochemical and Biophysical Research Communications,* 18, 48, 1965.
92. **Dunlap, P. V., and Greenberg, E. P.,** Control of *Vibrio fischeri* luminescence gene expression in *Escherichia coli* by cyclic AMP and cyclic AMP receptor protein, *Journal of Bacteriology,* 164, 45, 1985.
93. **Henry, J. P., and Michelson, A. M.,** Bioluminescence: physiological control and regulation at the molecular level, *Photochemistry and Photobiology,* 28, 293, 1978.
94. **Ne'eman, Z., Ulitzur, S., Branton, D., and Hastings, J. W.,** Membrane polypeptides co-induced with the bacterial bioluminescent system, *Journal of Biological Chemistry,* 252, 5150, 1977.
95. **Haun, M., Duran, N., and Cilento, G.,** Energy transfer from enzymatically generated triplet carbonyl compounds to the fluorescent state of flavins, *Biochemical and Biophysical Research Communications,* 81, 779, 1978.
96. **Kiel, J. L., and Erwin, D. N.,** Physiologic aging of mature porcine erythrocytes: effects of various metabolites, antimetabolites and physical stressors, *American Journal of Veterinary Research,* 47, 2155, 1986.
97. **Yubisui, T., Murakami, K., Takeshita, M., and Takano, T.,** Purification by hydrophobic chromatography of soluble cytochrome b₅ of human erythrocytes, *Biochimica et Biophysica Acta,* 936, 447, 1988.
98. **Strittmatter, P., Fleming, P., Connors, M., and Corcoran, D.,** Purification of cytochrome b₅, in *Methods in Enzymology, Biomembranes. Part C. Biological Oxidations: Microsomal, Cytochrome P450 and Other Hemoprotein Systems,* Vol. 2, Fleischer, S., and Packer, L., Eds., Academic Press, New York, 1978, 97.
99. **Hultquist, D. E., Dean, R. T., and Reed, D. W.,** Isolation and characterization of a green hemoprotein from human erythrocytes, *Journal of Biological Chemistry,* 251, 3927, 1976.
100. **Sono, M., Dawson, J. H., and Ikedo-Saito, M.,** Characterization of the spleen green hemoprotein with magnetic and natural circular dichroism spectroscopy: positive evidence for a myeloperoxidase-type active site, *Biochimica et Biophysica Acta,* 873, 62, 1986.

101. **Chee, P. P., and Lardy, H. A.,** Isolation from erythrocytes of a green hemoprotein with ferroactivator activity, *Journal of Biological Chemistry,* 256, 3865, 1981.

102. **Lewis, C. T., Seyer, J. M., and Carlson, G. M.,** Cysteine 288: an essential hyperactive thiol of cytosolic phosphenolpyruvate carboxykinase (GTP), *Journal of Biological Chemistry,* 264, 27, 1989.

103. **Williams, R. F.,** Preparation, Purification, and Characterization of Green Hemoprotein and Its Artificial Substrate, Diazoluminomelanin, Final Report on Task Order No. F33615–87–D-0626-0004, Armstrong Laboratory, U.S. Air force, San Antonio, TX, 1992.

104. **White, A., Handler, P., and Smith, E. L.,** *Principles of Biochemistry,* 4th Ed., McGraw-Hill, New York, 1968, chap. 18.

105. **Kiel, J. L., and Erwin, D. N.,** Microwave and thermal interactions with oxidative hemolysis, *Physiological Chemistry and Physics and Medical NMR,* 16, 317, 1984.

106. **Bentle, L. A., and Lardy, H. A.,** Interaction of anions and divalent metal ions with phosphoenolpyruvate carboxykinase, *Journal of Biological Chemistry,* 251, 2916, 1976.

107. **Kiel, J. L., McQueen, C., and Erwin, D. N.,** Interactions of methemoglobin and green hemoprotein in chemiluminescent gels, *Free Radical Biology and Medicine,* 8, 127, 1990.

108. **Stuehr, D. J., and Ikeda-Saito, M.,** Spectral characteristics of brain and macrophage nitric oxide synthases, *Journal of Biological Chemistry,* 267, 20547, 1992.

109. **Schmidt, H. H. H. W., Warner, T. D., Ishii, K., Sheng, H., and Murad, F.,** Insulin secretion from pancreatic B cells caused by L-arginine-derived nitrogen oxides, *Science,* 255, 721, 1992.

110. **Kiel, J. L.,** The Cytotoxic Activity of Peroxidase. Ph.D. Thesis. Texas Tech University Health Sciences Center, Lubbock, 1981.

111. **Kiel, J. L.,** U.S. Patent 4,486,408, 1984.

112. **Smith, D. C., and Klebanoff, S. J.,** A uterine fluid-mediated sperm-inhibitory system, *Biology of Reproduction,* 3, 229, 1970.

113. **Klebanoff, S. J., and Smith, D. C.,** The source of H_2O_2 for the uterine fluid-mediated sperm-inhibitory system, *Biology of Reproduction,* 3, 236, 1970.

114. **Lyttle, C. R., and De Sombre, E. R.,** Uterine peroxidase as a marker for estrogen action, *Proceedings of the National Academy of Science of the USA,* 74, 3162, 1977.

115. **Jellinck, P. H., and Newcombe, A.-M.,** Induction of peroxidase in the rat uterus under various endocrine conditions, *Journal of Endocrinology,* 74, 147, 1977.

116. **Olsen, R. L., and Little, C.,** Purification of rat uterine peroxidase, *Acta Chemica Scandinavica,* B35, 1, 1981.

117. **Lefkowitz, D. L., Lefkowitz, S. S., Wei, R.-Q., and Everse, J.,** Activation of macrophages with oxidative enzymes, in *Methods in Enzymology,* Vol. 132, Di Sabato, G., and Everse, J., Eds., Academic Press, Orlando, FL, 1986, 537.

118. **Kiel, J. L., and O'Brien, G. J.,** U.S. Patent 5,003,050, 1991.

119. **Prichard, P. M., and Cormier, M. J.,** Studies on the mechanism of the horseradish peroxidase catalyzed luminescent peroxidation of luminol, *Biochemical and Biophysical Research Communications,* 31, 131, 1968.

120. **Wright, J. R.,** NMR Characterization of Polymers Formed in Diazotizing Mixtures of Luminol and 3-Amino-L-tyrosine, Contract No. F49620–92-J-0191, Office of Scientific Research, U.S. Air Force, Bolling AFB, Washington, DC, 1993.

121. **Wright, J. R.,** NMR Characterization of Products Formed in Diazotizing Mixtures of Luminol and 3-Amino-L-tyrosine, Final Report of Grant 89-NL-258, Office of Scientific Research, U.S. Air Force, Bolling AFB, Washington, DC, 1991.

122. **Bruno, J. G., and Kiel, J. L.,** Luminol and diazoluminomelanin as indicators of HL-60 cell differentiation, *In Vitro: Cellular and Developmental Biology,* 29A, 737, 1993.

123. **Kiel, J. L.,** The ultimate biosensor, *Aviation, Space, and Environmental Medicine,* 65(5, Suppl.), A121, 1994.

124. **Merrifield, R. E.,** Magnetic effects on triplet exciton interactions, *Pure and Applied Chemistry,* 24, 481, 1971.
125. **McCauley, S. W., and Ruby, R. H.,** Delayed fluorescence induction in chloroplasts irradiance dependence, *Biochimica et Biophysica Acta,* 638, 268, 1981.
126. **Kepler, R. G., Caris, J. C., Avakian, P., and Abramson, E.,** Triplet excitons and delayed fluorescence in anthracene crystals, *Physical Review Letters,* 10, 400, 1963.
127. **Parker, C. A.,** Delayed fluorescence of 3:4-benzpyrene solutions, *Nature,* 200, 331, 1963.
128. **Drabent, R., and Laczko, G.,** Delayed fluorescence of the flavomononucleotide complex, *Acta Biochimica et Biophysica of the Acadimiae Scientiarum of Hungaricae,* 19, 265, 1984.
129. **Iwata, S., Tanaka, J., and Nagakura, S.,** Phosphorescence of the charge-transfer triplet states of some molecular complexes, *Journal of Chemical Physics,* 47, 2203, 1967.
130. **Tanaka, Y., and Azumi, T.,** Delayed fluorescence and triplet-triplet annihilation in a diplatinate (II) pyrophosphite complex, *Inorganic Chemistry,* 25, 247, 1986.
131. **Boxer, S. G., Chidsey, C. E. D., and Roelofs, M. G.,** Magnetic field effects on reaction yields in the solid state: an example from photosynthetic reaction centers, *Annual Reviews of Physical Chemistry,* 34, 389, 1983.
132. **Drabent, R., Mieloszyk, J., and Siodmiak, J.,** Delayed fluorescence and phosphorescence of flavomononucleotide stabilized by poly(vinyl alcohol) matrix, *Acta Biochimica et Biophysica Academiae Scientiarum Hungaricae,* 19, 259, 1984.
133. **Hideg, E., Scott, R. Q., and Inaba, H.,** High resolution emission spectra of one second delayed fluorescence from chloroplasts, *FEBS Letters,* 250, 275, 1989.
134. **Schenck, C. C., Blankenship, R. E., and Parson, W. W.,** Radical-pair decay kinetics, triplet yields and delayed fluorescence from bacterial reaction centers, *Biochimica et Biophysica Acta,* 680, 44, 1982.
135. **Conjeaud, H., Mathis, P., and Paillotin, G.,** Primary and secondary electron donors in photosystem II of chloroplasts. Rates of electron transfer and location in the membrane, *Biochimica et Biophysica Acta,* 546, 280, 1979.
136. **Filatova, J., McGinness, J., and Corry, P.,** Thermal and electronic contributions to switching in melanins, *Biopolymers,* 15, 2309, 1976.
137. **Crippa, P. R., Cristofoletti, V., and Romeo, N.,** A band model for melanin deduced from optical absorption and photoconductivity experiments, *Biochimica et Biophysica Acta,* 538, 164, 1978.
138. **Slawinska, D.,** Chemiluminescence and the formation of singlet oxygen in the oxidation of certain polyphenols and quinones, *Photochemistry and Photobiology,* 28, 453, 1978.
139. **Babcock, G. T., and Wikstrom, M.,** Oxygen activation and the conservation of energy in cell respiration, *Nature,* 356, 301, 1992.
140. **Cammack, R.,** Plug in a molecular diode, *Nature,* 356, 288, 1992.
141. **Sucheta, A., Ackrell, B. A. C., Cochran, B., and Armstrong, F. A.,** Diode-like behaviour of a mitochondrial electron-transport enzyme, *Nature,* 356, 361, 1992.
142. **Williams, R. J. P.,** Biology's wiring circuits, *Nature,* 355, 770, 1992.
143. **Rossi, F., Bellavite, P., and Papini, E.,** Respiratory response of phagocytes: terminal NADPH oxidase and the mechanism of its activation, in *Biochemistry of Macrophages,* Pitman, London, 1986, 172.
144. **Gleich, G. J., and Loegering, D. A.,** Immunobiology of eosinophils, *Annual Review of Immunology,* 2, 429, 1984.
145. **Clark, R. A., and Nauseef, W. M.,** Phagocyte oxidants and protective mechanisms, in *Cellular Antioxidant Defense Mechanisms,* Vol. II, Chow, C. K., Ed., CRC Press, Boca Raton, FL, 1988, chap. 23.
146. **Hadden, J. W., and Englard, A.,** Molecular aspects of macrophage activation and proliferation, in *Phagocytosis: Its Physiology and Pathology,* Kokubun, Y., and Kobayashi, N., Eds., University Park Press, Baltimore, 1978, 147.

147. **Nathan, C. F.,** Reactive oxygen intermediates in lysis of antibody-coated tumor cells, in *Macrophage-Mediated Antibody-Dependent Cellular Cytotoxicity,* Immunology Series, Vol. 21, Koren, H. S., Ed., Marcel Dekker, New York, 1983, chap. 15.

148. **Sbarra, A. J., Selvaraj, R. J., Paul, B. B., Mitchell, G. W., Jr., and Louis, F.,** Some newer insights of the peroxidase mediated antimicrobial system, in *Movement, Metabolism and Bactericidal Mechanisms of Phagocytes,* Rossi, F., Patriarca, P., and Romeo, D., Eds., Piccin Medical Books, Padua, Italy, 1977, 295.

149. **Johnston, R. B., Jr., and Lehmeyer, J. E.,** Neutrophil and monocyte oxidative metabolism in inflammation: stimulation by surface contact and suppression by anti-inflammation agents, in *Movement, Metabolism and Bactericidal Mechanisms of Phagocytes,* Rossi, F., Patriarca, P., and Romeo, D., Eds., Piccin Medical Books, Padua, Italy, 1977, 243.

150. **Rotilio, G., Brunori, M., Patriarca, P., and Dri, P.,** Studies on the role of myeloperoxidase in phagocytosis production of superoxide and EPR spectra, in *Movement, Metabolism and Bactericidal Mechanisms of Phagocytes,* Rossi, F., Patriarca, P., and Romeo, D., Eds., Piccin Medical Books, Padua, Italy, 1977, 207.

151. **Babior, B. M.,** The superoxide-forming enzyme responsible for the respiratory burst of neutrophils, in *Movement, Metabolism and Bactericidal Mechanisms of Phagocytes,* Rossi, F., Patriarca, P., and Romeo, D., Eds., Piccin Medical Books, Padua, Italy, 1977, 235.

152. **Root, R. K., and Metcalf, J. A.,** Superoxide and hydrogen peroxide formation by human granulocytes: inter-relationships and activation mechanisms. Studies with normal and cytochalasin-B treated cells, in *Movement, Metabolism and Bactericidal Mechanisms of Phagocytes,* Rossi, F., Patriarca, P., and Romeo, D., Eds., Piccin Medical Books, Padua, Italy, 1977, 185.

153. **Romeo, D., Zabucchi, G., and Rossi, F.,** Surface modulation of oxidative metabolism of polymorphonuclear leucocytes, in *Movement, Metabolism and Bactericidal Mechanisms of Phagocytes,* Rossi, F., Patriarca, P., and Romeo, D., Eds., Piccin Medical Books, Padua, Italy, 1977, 153.

154. **Patriarca, P., Cramer, R., and Dri, P.,** The present status of the subcellular localization of the NAD(P)H oxidase in polymorphonuclear leucocyte , in *Movement, Metabolism and Bactericidal Mechanisms of Phagocytes,* Rossi, F., Patriarca, P., and Romeo, D., Eds., Piccin Medical Books, Padua, Italy, 1977, 167.

155. **Parkos, C. A., Allen, R. A., Cochrane, C. G., and Jesaitis, A. J.,** Purified cytochrome b from human granulocyte plasma membrane is composed of two polypeptides with relative molecular weights of 91,000 and 22,000, *Journal of Clinical Investigations,* 80, 732, 1987.

156. **Capeillere-Blandin, C., Masson, A., and Descamps-Latsche, B.,** Molecular characteristics of cytochrome b_{558} isolated from human granulocytes, monocytes, and HL60 and U937 cells differentiated into monocytes/macrophages, *Biochimica et Biophysica Acta,* 1094, 55, 1991.

157. **Teahan, C., Rowe, P., Parker, P., Totty, N., and Segal, A. W.,** The X-linked chronic granulomatous diseaes gene codes for the beta-chain of cytochrome b_{-245}, *Nature,* 327, 720, 1987.

158. **Ding, A. H., and Nathan, C. F.,** The measurement of cytochrome b_{559} in polymorphonuclear leukocytes and macrophages in the presence of hemoglobin or mitochondrial cytochromes, *Analytical Biochemistry,* 175, 22, 1988.

159. **Dinauer, M. C., Orkin, S. H., Brown, R., Jesaitis, A. J., and Parkos, C. A.,** The glycoprotein encoded by the X-linked chronic granulomatous disease locus is a component of the neutrophil cytochrome b complex, *Nature,* 327, 717, 1987.

160. **Edwards, S. W., and Lloyd, D.,** The relationship between superoxide generation, cytochrome b and oxygen in activated neutrophils, *FEBS Letters,* 227, 39, 1988.

161. **Yamaguchi, T., Hayakawa, T., Kaneda, M., Kakinuma, K., and Yoshikawa, A.,** Purification and some properties of the small subunit of cytochrome b_{558} from human neutrophils, *Journal of Biological Chemistry,* 264, 112, 1989.

162. **Nunoi, H., Rotrosen, D., Gallin, J. I., and Malech, H. L.,** Two forms of autosomal chronic granulomatous disease lack distinct neutrophil cytosol factors, *Science,* 242, 1298, 1988.

163. **Nugent, J. H. A., Gratzer, W., and Segal, A. W.,** Identification of the haem-binding subunit of cytochrome b_{-245}, *Biochemical Journal,* 264, 921, 1989.

164. **Kleinberg, M. E., Rotrosen, D., and Malech, H. L.,** Asparagine-linked glycosylation of cytochrome b_{558} large subunit varies in different human phagocytic cells, *Journal of Immunology,* 143, 4152, 1989.

165. **Berton, G., Papini, E., Cassatella, M. A., Bellavite, P., and Rossi, F.,** Partial purification of the superoxide-generating system of macrophages. Possible association of the NADPH oxidase activity with a low-potential (−247 mV) cytochrome b, *Biochimica et Biophysica Acta,* 810, 164, 1985.

166. **Morel, F., and Vignais, P. V.,** Purification of cytochrome b_{558} from bovine polymorphonuclear neutrophils, *Biochemical and Biophysical Research Communications,* 149, 46, 1987.

167. **Volpp, B. D., Nauseef, W. M., and Clark, R. A.,** Two cytosolic neutrophil oxidase components absent in autosomal chronic granulomatous disease, *Science,* 242, 1295, 1988.

168. **Lutter, R., van Schaik, M. L. J., van Zwieten, R., Wever, R., Roos, D., and Hamers, M. N.,** Purification and partial characterization of the b-type cytochrome from human polymorphonuclear leukocytes, *Journal of Biological Chemistry,* 260, 2237, 1985.

169. **Lu, D. J., and Grinstein, S.,** ATP and guanine nucleotide dependence of neutrophil activation. Evidence for the involvement of two distinct GTP-binding proteins, *Journal of Biological Chemistry,* 265, 13721, 1990.

170. **Rotrosen, D., Yeung, C. L., Leto, T. L., Malech, H. L., and Kwang, C. H.,** Cytochrome b_{558}: the flavin-binding component of the phagocyte NADPH oxidase, *Science,* 256, 1459, 1992.

171. **Brozna, J. P., Hauff, N. F., Phillips, W. A., and Johnston, R. B., Jr.,** Activation of the respiratory burst in macrophages. Phosphorylation specifically associated with Fc receptor-mediated stimulation, *Journal of Immunology,* 141, 1642, 1988.

172. **Verhoeven, A. J., Bolscher, B. G. J. M., and Roos, D.,** The superoxide-generating enzyme in phagocytosis: physiology, protein composition, and mechanism of activation, in *Membrane Lipid Oxidation,* Vol. II, Vigo-Pelfrey, C., Ed., CRC Press, Boca Raton, FL, 1991, chap. 3.

173. **Bokoch, G. M., Quillian, L. A., Bohl, B. P., Jesaitis, A. J., and Quin, M. T.,** Inhibition of rap1A binding to cytochrome b_{558} of NADPH oxidase by phosphorylation of rap1A, *Science,* 254, 1794, 1991.

174. **Knaus, U. G., Heyworth, P. G., Evans, T., Curnutte, J. T., and Bokoch, G. M.,** Regulation of phagocyte oxygen radical production by the GTP-binding protein rac2, *Science,* 254, 1512, 1991.

175. **Lomax, K. J., Leto, T. L., Nunoi, H., Gallin, J. I., and Malech, H. L.,** Recombinant 47-kilodalton cytosol factor restores NADPH oxidase in chronic granulomatous disease, *Science,* 245, 409, 1989.

176. **Smith, C. D., Cox, C. C., and Snyderman, R.,** Receptor-coupled activation of phosphoinositide-specific phospholipase C by an N protein, *Science,* 232, 97, 1986.

177. **Struhar, D., and Harbeck, R. J.,** Inhibition of induced acute lung edema by a novel protein kinase C inhibitor, *FASEB Journal,* 1, 116, 1987.

178. **Adams, D. O.,** Macrophage activation and secretion, *Federation Proceedings,* 41, 2193, 1982.

179. **Karnovsky, M. L., Badwey, J. A., Curnutte, J. T., Robinson, J. M., Berde, C. B., and Karnovsky, M. J.**, The excitable membranes of phagocytic leukocytes, in *Information and Energy Transduction in Biological Membranes,* Alan R. Liss, New York, 1984, 347.

180. **Putney, J. W., Jr.**, Calcium and phosphoinositides in physiological and pathological mechanisms of cell regulation, in *Lilly Research Laboratories Symposium/Molecular Toxicology,* May 1990, 56.

181. **McPhail, L. C., Clayton, C. C., and Snyderman, R.**, A potential second messenger role for unsaturated fatty acids: activation of Ca^{2+}-dependent protein kinase, *Science,* 224, 622, 1984.

182. **Majerus, P. W., Connally, T. M., Deckmyn, H., Ross, T. S., Bross, T. E., Ishii, H., Bansal, V. S., and Wilson, D. B.**, The metabolism of phosphoinositide-derived messenger molecules, *Science,* 234, 1519, 1986.

183. **Murray, H. W., Spitalny, G. L., and Nathan, C. F.**, Activation of mouse peritoneal macrophages *in vitro* and *in vivo* by interferon-gamma, *Journal of Immunology,* 134, 1619, 1985.

184. **Hirai, K.-I., Moriguchi, K., and Wang, G.-Y.**, Human neutrophils produce free radicals from the cell-zymosan interface during phagocytosis and from the whole plasma membrane when stimulated with calcium ionophore A23187, *Experimental Cell Research,* 194, 19, 1991.

185. **Yamamoto, H., and Suzuki, T.**, Prostaglandin E_2-induced activation of adenosine 3'-5' cyclic monophosphate-dependent protein kinases of a murine macrophage-like cell line (P388D$_1$), *Journal of Immunology,* 139, 3416, 1987.

186. **Weinstein, S. L., Gold, M. R., and De Franco, A. L.**, Bacterial lipopolysaccharide stimulates protein tyrosine phosphorylation in macrophages, *Proceedings of the National Academy of Sciences of the USA,* 88, 4148, 1991.

187. **Engler, R.**, Consequences of activation and adenosine-mediated inhibition of granulocytes during myocardial ischemia, *Federation Proceedings,* 46, 2407, 1987.

188. **Naccache, P. H., Molski, T. F. P., Borgeat, P., White, J. R., and Sha'afi, R. I.**, Phorbol esters inhibit the fmet-leu-phe- and leukotriene B$_4$-stimulated calcium mobilization and enzyme secretion in rabbit neutrophils, *Journal of Biological Chemistry,* 260, 2125, 1985.

189. **Odin, J. A., Edberg, J. C., Painter, C. J., Kimberly, R. P., and Unkeless, J. C.**, Regulation of phagocytosis and $[Ca^{2+}]_i$ flux by distinct regions c ˙ an Fc receptor, *Science,* 254, 1785, 1991.

190. **Johnston, Jr., R. B., and Kitagawa, S.**, Molecular basis for the enhanced respiratory burst of activated macrophages, *Federation Proceedings,* 44, 2927, 1985.

191. **Nishizuka, Y.**, Studies and perspectives of protein kinase C, *Science,* 233, 305, 1986.

192. **Simchowitz, L., Foy, M. A., and Cragoe, E. J., Jr.**, A role for Na^+/Ca^{2+} exchange in the generation of superoxide radicals by human neutrophils, *Journal of Biological Chemistry,* 265, 13449, 1990.

193. **Rossi, F., Bellavite, P., and Papini, E.**, Respiratory response of phagocytes: terminal NADPH oxidase and the mechanisms of its activation, in *Biochemistry of Macrophages,* Pitman, London, 1986, 172.

194. **Tsunawaki, S., and Nathan, C.**, Release of arachidonate and reduction of oxygen. Independent metabolic bursts of the mouse peritoneal macrophage, *Journal of Biological Chemistry,* 261, 11563, 1986.

195. **Parker, J., Daniel, L. W., and Moseley, W.**, Evidence of protein kinase C involvement in phorbol diester-stimulated arachidonic acid release and prostaglandin synthesis, *Journal of Biological Chemistry,* 262, 5385, 1987.

196. **Aderem, A. A., and Cohn, Z. A.**, Bacterial lipopolysaccharides modify signal transduction in the arachidonic acid cascade in macrophages, in *Biochemistry of Macrophages,* Pitman, London, 1986, 196.

197. **Bromberg, Y., and Pick, E.,** Activation of NADPH-dependent superoxide production in a cell-free system by sodium dodecyl sulfate, *Journal of Biological Chemistry,* 260, 13539, 1985.

198. **Lehnmann, V., Benninghoff, B., and Droge, W.,** Tumor necrosis factor-induced activation of peritoneal macrophages is regulated by prostaglandin E_2 and cAMP, *Journal of Immunology,* 141, 587, 1988.

199. **Tsai, M.- H., Yu, C.- L., Wei, F.- S., and Stacey, D. W.,** The effect of GTPase activating protein upon ras is inhibited by mitogenically responsive lipids, *Science,* 243, 522, 1989.

200. **Simons, J. M., 't Hart, B. A., Ie Vai Ching, T. R. A. M., Van Dijk, H., and Labadie, R. P.,** Metabolic activation of natural phenols into selective oxidative burst agonists by activated human neutrophils, *Free Radical Biology & Medicine,* 8, 251, 1990.

201. **'t Hart, B. A., Simmons, J. M., Rijkers, G. T., Hooguliet, J. C., Van Dijk, H., and Labadie, R. P.,** Reaction products of 1-naphthol with reactive oxygen species prevent NADPH oxidase activation in activated human neutrophils, but leave phagocytosis intact, *Free Radical Biology & Medicine,* 8, 241, 1990.

202. **Yamaguchi, T., Hayakawa, T., Kaneda, M., Kakinuma, K., and Yoshikawa, A.,** Purification and some properties of the small subunit of cytochrome b_{558} from human neutrophils, *Journal of Biological Chemistry,* 264, 112, 1989.

203. **Iizuka, T., Kanegasaki, S., Makino, R., Tanaka, T., and Ishimura, Y.,** Pyridine and imidazole reversibly inhibit the respiratory burst in porcine and human neutrophils: evidence for the involvement of cytochrome b_{558} in the reaction, *Biochemical and Biophysical Research Communications,* 130, 621, 1985.

204. **Engler, R.,** Granulocytes and oxidative injury in myocardial ischemia and reperfusion. Introduction, *Federation Proceedings,* 46, 2395, 1987.

205. **Harvath, L., Amirault, H. J., and Andersen, B. R.,** Chemiluminescence of human and canine polymorphonuclear leukocytes in the absence of phagocytosis, *Journal of Clinical Investigations,* 61, 1145, 1978.

206. **Fox, I. H., and Kelley, W. N.,** The role of adenosine and 2′-deoxyadenosine in mammalian cells, *Annual Review of Biochemistry,* 47, 655, 1978.

207. **Kvietys, P. R., Smith, S. M., Grisham, M. B., and Manci, E. A.,** 5-Aminosalicylic acid protects against ischemia/reperfusion-induced gastric bleeding in the rat, *Gastroenterology,* 94, 733, 1988.

208. **Berton, G., Papini, E., Cassatella, M. A., Bellvite, P., and Rossi, F.,** Partial purification of the superoxide-generating system of macrophages. Possible association of the NADPH oxidase activity with a low-potential (-247 mV) cytochrome b, *Biochimica et Biophysica Act,* 810, 164, 1985.

209. **Green, T. R., and Pratt, K. L.,** Purification of the solubilized NADPH:O_2 oxidoreductase of human neutrophils, *Journal of Biological Chemistry,* 263, 5617, 1988.

210. **Morehouse, L. A., and Aust, S. D.,** Microsomal oxygen radical generation—relationship to the initiation of lipid peroxidation, in *Cellular Antioxidant Defense Mechanisms,* Vol. I, Chow, C. K., Ed., CRC Press, Boca Raton, FL, 1988, chap. 1.

211. **Parkos, C. A., Dinauer, M. C., Walker, L. E., Allen, R. A., Jesaitis, A. J., and Orkin, S. H.,** Primary structure and unique expression of the 22-kilodalton light chain of human neutrophil cytochrome b, *Proceedings of the National Academy of Sciences of the USA,* 85, 3319, 1988.

212. **Seifert, R., Jungblut, P., and Schultz, G.,** Differential expression of cytosolic activation factors for NADPH oxidase in HL-60 leukemic cells, *Biochemical and Biophysical Research Communications,* 161, 1109, 1989.

213. **Levy, R., Rotrosen, D., Nagauker, O., Leto, T. L., and Malech, H. L.,** Induction of the respiratory burst in HL-60 cells. correlation of function and protein expression, *Journal of Immunology,* 145, 2595, 1990.

214. **Thompson, B. Y., Sivam, G., Britigan, B. E., Rosen, G. M., and Cohen, M. S.**, Oxygen metabolism of the HL-60 cell line: comparison of the effects of monocytoid and neutrophilic differentiation, *Journal of Leukocyte Biology,* 43, 140, 1988.

215. **Roberts, P. J., Cross, A. R., Jones, O. T. G., and Segal, A. W.**, Development of cytochrome b and an active oxidase system in associations with maturation of a human promyelocytic (HL-60) cell line, *Journal of Cell Biology,* 95, 720, 1982.

216. **Pick, E., and Gadba, R.**, Certain lymphoid cells contain the membrane-associated component of the phagocyte-specific NADPH oxidase, *Journal of Immunology,* 140, 1611, 1988.

217. **Maly, F. E., Cross, A. R., Jones, O. T. G., Wolf-Vorbeck, G., Walker, C., Dahinden, C. A., and De Weck, A. L.**, The superoxide generating system of B cell lines. Structural homology with the phagocytic oxidase and triggering via surface Ig, *Journal of Immunology,* 140, 2334, 1988.

218. **Kiel, J. L., Parker, J. E., Alls, J. L., and Pruett, S. B.**, The cellular stress transponder: mediator of electromagnetic effects or artifacts?, *Nanobiology,* 1, 491, 1992.

219. **Storz, G., Tartaglia, L. A., and Ames. B. N.**, Transcriptional regulator of oxidative stress-inducible genes: direct activation by oxidation, *Science,* 248, 189, 1990.

220. **Schlesinger, M. J.**, Heat shock proteins, *Journal of Biological Chemistry,* 265, 12111, 1990.

221. **Goetz, M. B., and Proctor, R. A.**, Effect of endotoxin on neutrophil oxidative metabolism and chemiluminescence, in *Cellular Chemiluminescence,* Vol. II, van Dyke, K., and Castranova, V., Eds., CRC Press, Boca Raton, FL, 1987, chap. 6.

222. **Buchmuller, Y., and Mauel, J.**, Studies on the mechanisms of macrophage activation: possible involvement of oxygen metabolites in killing of *Leishmania ensiettii* by activated mouse macrophages, *Journal of the Reticuloendothelial Society,* 29, 181, 1981.

223. **Torres, M., Auclair, C., and Hakin, J.**, Protein-mediated hydroxyl radical generation—the primary event in NADH oxidation and oxygen reduction by the granule rich fraction of human resting leukocytes, *Biochemical and Biophysical Research Communications,* 88, 1003, 1979.

224. **Descamps-Latscha, B., Golub, R. M., Nguyen, A. T., and Feuillet-Fieux, M.-N.**, Monoclonal antibodies against T cell differentiation antigens initiate stimulation of monocyte/macrophage oxidative metabolism, *Journal of Immunology,* 131, 2500, 1983.

225. **Nakagawa, A., De Santis, N. M., Nogueira, N., and Nathan, C. F.**, Lymphokines enhance the capacity of human monocytes to secrete reactive oxygen intermediates, *Journal of Clinical Investigations,* 70, 1042, 1982.

226. **Wolpe, S. D., and Cerami, A.**, Macrophage inflammatory proteins 1 and 2: members of a novel superfamily of cytokines, *FASEB Journal,* 3, 2565, 1989.

227. **Goldman, D. W., Gifford, L. A., Marotti, T., Koo, C. H., and Goetzl, E. J.**, Molecular and cellular properties of human polymorphonuclear leukocyte receptors for leukotriene B$_4$, *Federation Proceedings,* 46, 200, 1987.

228. **Stuehr, D. J., and Marletta, M. A.**, Induction of nitrite/nitrate synthesis in murine macrophages by BCG infection, lymphokines, or interferon-gamma, *Journal of Immunology,* 139, 518, 1987.

229. **Nathan, C. F., Murray, H. W., Wiedbe, M. E., and Rubin, B. Y.**, Identification of interferon-gamma as the lymphokine that activates human macrophage oxidative metabolism and antimicrobial activity, *Journal of Experimental Medicine,* 158, 670, 1983.

230. **Kaplan, G., Nathan, C. F., Gandhi, R., Horwitz, M. A., Levio, W. R., and Cohn, Z. A.**, Effect of recombinant interferon-gamma on hydrogen peroxide-releasing capacity of monocyte-derived macrophages from patients with lepromatous leprosy, *Journal of Immunology,* 137, 983, 1986.

231. **Iyengar, R., Stuehr, D. J., and Marletta, M. A.**, Macrophage synthesis of nitrite, nitrate and *N*-nitrosamines: precursors and role of the respiratory burst, *Proceedings of the National Academy of Sciences of the USA,* 84, 6369, 1987.

232. **Coyle, J. T., and Puttfarcken, P.,** Oxidative stress, glutamate, and neurodegenerative disorders, *Science,* 262, 689, 1993.
233. **Stamler. J. S., Singel, D. J., and Loscalzo, J.,** Biochemistry of nitric oxide and its redox-activated forms, *Science,* 258, 1898, 1992.
234. **Bredt, D. S., and Snyder, S. H.,** Nitric oxide mediates glutamate-linked enhancement of cGMP levels in the cerebellum, *Proceedings of the National Academy of Sciences of the USA,* 86, 9030, 1989.
235. **Meltzer, M. S., Occhionero, M., and Ruco, L. P.,** Macrophage activation for tumor cytotoxicity: regulatory mechanisms for induction and control of cytotoxic activity, *Federation Proceedings,* 41, 2198, 1982.
236. **Nathan, C. F.,** Secretion of oxygen intermediates: role in effector functions of activated macrophages, *Federation Proceedings,* 41, 2206, 1982.
237. **Johnston, R. B., Jr., Keele, B. B., Jr., Misra, H. P., Lehmeyer, J. E., Webb, L. S., Baehner, R. L., and Rajagopalan, K. V.,** The role of superoxide anion generation in phagocytic bactericidal activity, *Journal of Clinical Investigations,* 55, 1357, 1975.
238. **Aust, S. D., and Miller, D. M.,** Role of iron in oxygen radical generation and reactions, *Lilly Research Laboratories Symposium/Molecular Toxicology,* May, 29, 1990.
239. **Nathan, C. F., Arrick, B. A., Murray, H. W., De Santis, N. M., and Cohn, Z. A.,** Tumor cell antioxidant defenses. Inhibition of the glutathione redox cycle enhances macrophage-mediated cytolysis, *Journal of Experimental Medicine,* 153, 766, 1980.
240. **Hassett, D. J., and Cohen, M. S.,** Bacterial adaptation to oxidative stress: implications for pathogenesis and interaction with phagocytic cells, *FASEB Journal,* 3, 2574, 1989.
241. **Tsunawaki, S., Sporn, M., Ding, A., and Nathan, C.,** Deactivation of macrophages by transforming growth factor-beta, *Nature,* 334, 260, 1988.
242. **Srimal, S., and Nathan, C.,** Purification of macrophage deactivating factor, *Journal of Experimental Medicine,* 171, 1347, 1990.
243. **Harris, E. D.,** Regulation of antioxidant enzymes, *FASEB Journal,* 6, 2675, 1992.
244. **Tsunawaki, S., Sporn, M., and Nathan, C.,** Comparison of transforming growth factor-beta and a macrophage-deactivating polypeptide from tumor cells. Differences in antigenicity and mechanism of action, *Journal of Immunology,* 142, 3462, 1989.
245. **Szuro-Sudol, A., and Nathan, C.,** Suppression of macrophage oxidative metabolism by products of malignant and nonmalignant cells, *Journal of Experimental Medicine,* 156, 945, 1982.
246. **DeChatelet, L. R., Migler, R. A., Shirley, P. S., Bass, D. A., and McCall, C. E.,** Enzymes of oxidative metabolism in the human eosinophil, *Proceedings of the Society for Experimental Biology and Medicine,* 158, 537, 1978.
247. **Klebanoff, S. J., Durak, D. T., Rosen, H., and Clark, R. A.,** Functional studies on human peritoneal eosinophils, *Infection and Immunity,* 17, 167, 1977.
248. **DeChatelet, L. R., Shirley, P. S., McPhail, L. C., Huntley, C. C., Muss, H. B., and Bass, D. A.,** Oxidative metabolism of the human eosinophil, *Blood,* 50, 525, 1977.
249. **Takenaka, T., Okuda, M., Kawabori, S., and Kubo, K.,** Extracellular release of peroxidase from eosinophils by interaction with immune complexes, *Clinical Experimental Immunology,* 28, 56, 1977.
250. **Weiss, S. J., Test, S. T., Eckmann, C. M., Roos, D., and Regiani, S.,** Brominating oxidants generated by human eosinophils, *Science,* 234, 200, 1986.
251. **Henderson, W. R., Chi, E. Y., and Klebanoff, S. J.,** Eosinophil peroxidase-induced mast cell secretion, *Journal of Experimental Medicine,* 152, 265, 1980.
252. **Oto, Y. H., Kurozumi, S., Azuma, A., Komoriya, K., and Ohmori, H.,** Xanthine oxidase-induced histamine release from isolated rat peritoneal mast cells: involvement of hydrogen peroxide, *Biochemical Pharmacology,* 28, 333, 1979.
253. **Christie, K. N., and Stoward, P. J.,** Endogenous peroxidase in mast cells localized with a semipermeable membrane technique, *Histochemical Journal,* 10, 425, 1978.

254. **Henderson, W. R., Chi, E. Y., Jong, E. C., and Klebanoff, S. J.,** Mast cell-mediated tumor-cell cytotoxicity, *Journal of Experimental Medicine,* 153, 520, 1981.

255. **Nathan, C. F., and Tsunawaki, S.,** Secretion of toxic oxygen products by macrophages: regulatory cytokines and their effects on the oxidase, in *Biochemistry of Macrophages,* Pitman, London, 1986, 211.

256. **Tsunawaki, S., and Nathan, C. F.,** Enzymatic basis of macrophage activation. Kinetic analysis of superoxide production in lysates of resident and activated mouse peritoneal macrophages and granulocytes, *Journal of Biological Chemistry,* 259, 4305, 1984.

257. **Davila, D. R., Edwards, C. K., III, Arkins, S., Simon, J., and Kelley, K. W.,** Interferon-gamma-induced priming for secretion of superoxide anion and tumor necrosis factor-alpha declines in macrophages from aged rats, *FASEB Journal,* 4, 2906, 1990.

258. **Nakagawara, A., Nathan, C. F., and Cohn, Z. A.,** Hydrogen peroxide metabolism in human monocytes during differentiation in vitro, *Journal of Clinical Investigations,* 68, 1243, 1981.

259. **Najjar, V. A.,** Tuftsin (thr-lys-pro-arg): a natural activator of phagocytic cells with antibacterial and antineoplastic activity, in *New Approaches to Disease Intervention,* Forrence, P. F., Ed., Academic Press, New York, 1985, chap. 7.

260. **Adams, D. O., Johnson, W. J., and Marino, P. A.,** Mechanisms of target recognition and destruction in macrophage-mediated tumor cytotoxicity, *Federation Proceedings,* 41, 2212, 1982.

261. **Sarnet, J. M., and Friedman, M.,** Effect of ozone on platelet activating factor metabolism in phorbol-differentiated HL-60 cells, *Toxicology and Applied Pharmacology,* 117, 19, 1992.

262. **Williams, Z., Hertogs, C. F., and Pluznik, D. H.,** Use of mice tolerant to lipopolysaccharide to demonstrate requirement of cooperation between macrophages and lymphocytes to generate lipopolysaccharide-induced colony-stimulating factor *in vivo, Infection and Immunity,* 41, 1, 1983.

263. **Old, L. J.,** Tumor necrosis factor (TNF), *Science,* 230, 630, 1985.

264. **Abina, J. E., Cui, S., Mateo, R. B., and Reichner, J. S.,** Nitric oxide-mediated apoptosis in murine peritoneal macrophages, *Journal of Immunology,* 150, 5080, 1993.

265. **Pruett, S. B., and Kiel, J. L.,** Quantitative aspects of the feeder cell phenomenon: mechanistic implications, *Biochemical and Biophysical Research Communications,* 150, 1037, 1988.

266. **Russell, J. H.,** Phorbol-ester stimulated lysis of weak and nonspecific target cells by cytotoxic T lymphocytes, *Journal of Immunology,* 136, 23, 1986.

267. **Marx, J. L.,** How killer cells kill their targets, *Science,* 231, 1367, 1986.

268. **Wong, L. S., and Kiel, J. L.,** Effects of antigens on mononuclear leukocyte chemiluminescence, *Immunological Investigations,* 14, 503, 1985.

269. **Koenig, S., and Hoffman, M. K.,** Bacterial lipopolysaccharide activates suppressor B lymphocytes, *Proceedings of the National Academy of Sciences of the USA,* 76, 4608, 1979.

270. **Lopez, D. M., Blomberg, B. B., Padmanabhan, R. R., and Bourguignon, L. Y. W.,** Nuclear disintegration of target cells by killer B lymphocytes from tumor-bearing mice, *FASEB Journal,* 3, 37, 1989.

271. **Nathan, C. F., Mercer-Smith, J. A., De Santis, N. M., and Palladino, M. A.,** Role of oxygen in T cell-mediated cytolysis, *Journal of Immunology,* 129, 2164, 1982.

272. **Kyner, D., Christman, J. K., and Acs, G.,** Effect of tumor promoters on induction of aryl hydrocarbon hydroxylase in human lymphocytes, *Enzymes,* 25, 322, 1980.

273. **Fan, J., and Bonavida, B.,** Studies on the induction and expression of T cell-mediated immunity. XIV. Antigen-nonspecific oxidation-dependent cellular cytotoxicity (ODCC) mediated by sodium periodate oxidation of cytotoxic T lymphocytes, *Journal of Immunology,* 131, 1426, 1983.

274. **Wong, L. S., and Kiel, J. L.,** Anamnestic chemiluminescence of murine spleen cells, *Immunological Communications,* 13, 285, 1984.

275. **Steffen, M., Ottman, O. G., and Moore, M. A. S.,** Simultaneous production of tumor necrosis factor-alpha and lymphotoxin by normal T cells after induction with IL-2 and anti-T_3, *Journal of Immunology,* 140, 2621, 1988.

276. **Gately, C. L., Wahl, S. M., and Oppenheim, J. J.,** Characterization of hydrogen peroxide-potentiating factor. A lymphokine that increases the capacity of human monocytes and monocyte-like cell lines to produce hydrogen peroxide, *Journal of Immunology,* 131, 2853, 1983.

277. **Rossi, F., Della Bianca, V., and Bellavite, P.,** Inhibition of the respiratory burst and of phagocytosis by nordihydroguataretic acid in neutrophils, *FEBS Letters,* 127, 183, 1981.

278. **Bonney, R. J., Opas, E. E., and Humes, J. L.,** Lipoxygenase pathways of macrophages, *Federation Proceedings,* 44, 2933, 1985.

279. **Chen, C., and Hirsch, J. G.,** The effects of mercaptoethanol and of peritoneal macrophages on the antibody-forming capacity of nonadherent mouse spleen cells *in vitro, Journal of Experimental Medicine,* 136, 604, 1972.

280. **Strauss, R. R., Friedman, H., Milk, L., and Zayon, G.,** Relationship between antigenic stimulation and increased splenic peroxidase levels during the immune response, *Infection and Immunity,* 15, 197, 1977.

281. **Oda, T., Akaike, T., Hamamoto, T., Suzuki, F., Hirano, T., and Maeda, H.,** Oxygen radicals in influenza-induced pathogenesis and treatment with pyran polymer-conjugated SOD, *Science,* 244, 974, 1989.

282. **Manzella, J. P., and Roberts, N. J., Jr.,** Human macrophage and lymphocyte responses to mitogen stimulation after exposure to influenza virus, ascorbic acid, and hyperthermia, *Journal of Immunology,* 123, 1940, 1979.

283. **Smail, E. H., Melnick, D. A., Ruggeri, R., and Diamond, R. D.,** A novel natural inhibitor from *Candida albicans* hyphae causing dissociation of the respiratory burst response to chemotactic peptides from other post-activation events, *Journal of Immunology,* 140, 3893, 1988.

284. **Bodwey, J. A., and Karnovsky, M. L.,** Production of superoxide and hydrogen peroxide by an NADH-oxidase in guinea pig polymorphonuclear leukocytes, *Journal of Biological Chemistry,* 254, 11530, 1979.

285. **Gartner, S., Markovits, P., Markovitz, D. M., Kaplan, M. H., Gallo, R. C., and Popovic, M.,** The role of mononuclear phagocytes in HTLV-III/LAV infection, *Science,* 233, 215, 1986.

286. **Manev, H., Costa, E., Wroblewski, J. T., and Guidotti, A.,** Abusive stimulation of excitatory amino acid receptors: a strategy to limit neurotoxicity, *FASEB Journal,* 4, 2789, 1990.

287. **McCord, J. M.,** Oxygen-derived radicals: a link between reperfusion injury and inflammation, *Federation Proceedings,* 46, 2402, 1987.

288. **Kontos, H. A.,** Oxygen radicals in cerebral vascular responses, in *Physiology of Oxygen Radicals,* Taylor, A. E., Matalon, S., and Ward, P., Eds., American Physiological Society, Bethesda, MD, 1986, chap. 16.

289. **Kontos, H. A., Wei, E. P., Christman, C. W., Levasseur, J. E., Povlishock, J. T., and Ellis, E. F.,** Free oxygen radicals in cerebral vascular responses, *The Physiologist,* 26, 165, 1983.

290. **Granger, D. N., and Parks, D. A.,** Role of oxygen radicals in the pathogenesis of intestinal ischemia, *The Physiologist,* 26, 159, 1983.

291. **Colton, C. A., and Gilbert, D. L.,** An endogenous source of the superoxide anion in the central nervous system, in *Oxygen Radicals in Biology and Medicine,* Simic, M. G., Taylor, K. A., Ward, J. F., and von Sonntag, C., Eds., Plenum Press, New York, 1988, 1005.

292. **Birkle, D. L., and Bazan, N. G.**, Metabolism of arachidonic acid in the central nervous system. The enzymatic cyclooxygenation and lipoxygenation of arachidonic acid in the mammalian retina, in *Phospholipids in the Nervous System: Physiological Roles,* Vol. 2, Harrocks, L. A., Ed., Raven Press, New York, 1985, 193.

293. **Friedl, H. P., Till, G. O., Ryan, U. S., and Ward, P. A.**, Mediator-induced activation of xanthine oxidase in endothelial cells, *FASEB Journal,* 3, 2512, 1989.

294. **Guilian, D., Woodward, J., Young, D. G., Krebs, J. F., and Lachman, L. B.**, Interleukin-1 injected into mammalian brain stimulates astrogliosis and neovascularization, *Journal of Neuroscience,* 8, 2485, 1988.

295. **Spina, M. B., and Cohen, G.**, Hydrogen peroxide in dopamine neurons, in *Oxygen Radicals in Biology and Medicine,* Simic, M. G., Taylor, K. A., Ward, J. F., and von Sonntag, C., Eds., Plenum Press, New York, 1988, 1011.

296. **Olney, J. W., Zorumski, C., Price, M. T., and Labruyere, J.**, L-Cysteine, a bicarbonate-sensitive endogenous excitotoxin, *Science,* 248, 596, 1990.

297. **Edwards, C. K., III, Ghiasuddin, S. M., Schepper, J. M., Yunger, L. M., and Kelley, K. W.**, A newly defined property of somatotropin: priming of macrophages for production of superoxide anion, *Science,* 239, 769, 1988.

298. **Hickey, W. F., and Kimura, H.**, Perivascular microglial cells of the CNS are bone marrow-derived and present antigen *in vivo, Science,* 239, 290, 1988.

299. **Guilian, D., and Lachman, L. B.**, Interleukin-1 stimulation of astroglial proliferation after brain injury, *Science,* 228, 497, 1985.

300. **Anonymous**, IL-1 and the glial response to injury, *Neuroscience Facts,* 2, 4, 1991.

301. **Frei, K., Bodmer, S., Schwerdel, C., and Fontana, A.**, Astrocyte-derived interleukin 3 as a growth factor for microglia cells and peritoneal macrophages, *Journal of Immunology,* 137, 3521, 1986.

302. **Huang, L., Privalle, C. T., Serafin, D., and Klitzman, B.**, Increased survival of skin flaps by scavengers of superoxide radical, *FASEB Journal,* 1, 129, 1987.

303. **Dinarello, C. A.**, Biology of interleukin 1, *FASEB Journal,* 2, 108, 1988.

304. **Edelson, R. L., and Fink, J. M.**, The immunologic function of skin, *Scientific American,* 252, 46, 1985.

305. **Norris, D. A., and Lee, L. A.**, Antibody-dependent cellular cytotoxicity and skin disease, *Journal of Investigative Dermatology,* 85, 1655, 1985.

306. **Umetsu, D. T., Katzen, D., Jabara, H. H., and Geha, R. S.**, Antigen presentation by human dermal fibroblasts: activation of resting T lymphocytes, *Journal of Immunology,* 136, 440, 1986.

307. **Heck, D. E., Laskin, D. L., Gardner, C. R., and Laskin, J. D.**, Role of nitric oxide in chemical induced skin injury, *The Toxicologist,* 13, 183, Abstr. 662, 1993.

308. **Nathan, C. F.**, Secretory products of macrophages, *Journal of Clinical Investigations,* 79, 319, 1987.

309. **Nathan, C. F.**, Secretion of oxygen intermediates: role in effector functions of activated macrophages, *Federation Proceedings,* 41, 2206, 1982.

310. **Wong, P. K., Hampton, M. J., and Floyd, R. A.**, Evidence for lipoxygenase-peroxidase activation of *N*-hydroxy-2-acetylaminofluorene by rat mammary gland parenchymal cells, in *Prostaglandins and Cancer: First International Conference,* Allan R. Liss, New York, 1982, 167.

311. **Lyttle, C. R., and De Sombre, E. R.**, Generality of oestrogen stimulation of peroxidase activity in growth responsive tissues, *Nature,* 268, 337, 1977.

312. **Anderson, W. A., De Sombre, E. R., and Kong, Y.-H.**, Estrogen-progesterone antagonism with respect to specific marker protein synthesis and growth by the uterine endometrium, *Biology of Reproduction,* 16, 409, 1977.

313. **McLachlan, J. A.**, Role of metabolism in the activity and toxicity of xenobiotic estrogens. Introduction, *Federation Proceedings,* 46, 1854, 1987.

314. **Metzler, M.**, Metabolic activation of xenobiotic stilbene estrogens, *Federation Proceedings*, 46, 1855, 1987.

315. **Kiel, J. L.**, U.S. Patent 4,766,150, 1988.

316. **Dimitrijevic, L., Aussel, C., Mucchielli, A., and Masseyeff, R.**, Purification and characterization of an estrogen-binding peroxidase from human fetuses, *Biochimie*, 61, 535, 1979.

317. **Foerder, C. A., Klebanoff, S. J., and Shapiro, B. M.**, Hydrogen peroxide production, chemiluminescence, and the respiratory burst of fertilization: interrelated events in early sea urchin development, *Proceedings of the National Academy of Sciences of the USA*, 75, 3183, 1978.

318. **Klebanoff, S. J., Foerder, C. A., Eddy, E. M., and Shapiro, B. M.**, Metabolic similarities between fertilization and phagocytosis. Conservation of a peroxidative mechanism, *Journal of Experimental Medicine*, 149, 938, 1979.

319. **Vernon, M., Foerder, C., Eddy, E. M., and Shapiro, B. M.**, Sequential biochemical and morphological events during assembly of the fertilization membrane of the sea urchin, *Cell*, 10, 321, 1977.

320. **Shapiro, B. M.**, The control of oxidant stress at fertilization, *Science*, 252, 533, 1991.

321. **Billiar, T. R., Curran, R. D., West, M. A., Hofmann, K., and Simmons, R. L.**, Kupffer cell cytotoxicity to hepatocytes in coculture requires L-arginine, *Archives of Surgery*, 124, 1416, 1989.

322. **Lepay, D. A., Nathan, C. F., Steinman, R. M., Murray, H. W., and Cohn, Z. A.**, Murine Kupffer cells. Mononuclear phagocytes deficient in the generation of reactive oxygen intermediates, *Journal of Experimental Medicine*, 161, 1079, 1985.

323. **Edwards, M. J., Keller, B. J., Kauffman, F. C., and Thurman, R. G.**, The involvement of Kupffer cells in carbon tetrachloride toxicity, *Toxicology and Applied Pharmacology*, 119, 275, 1993.

324. **Billiar, T. R., Curran, R. D., Harbrecht, B. G., Stuehr, D. J., Demetris, A. J., and Simmons, R. L.**, Modulation of nitrogen oxide synthesis in vivo: N^G-monomethyl-L-arginine inhibits endotoxin-induced nitrite/nitrate biosynthesis while promoting hepatic damage, *Journal of Leukocyte Biology*, 48, 565, 1990.

325. **Anonymous,** Fidia Research Foundation symposium: "excitatory amino acids 1990", *Neuroscience Facts*, 1, 3, 1990.

326. **Weddle, C. C., Hornbrook, R., and McCay, P. B.**, Lipid peroxidation and alteration of membrane lipids in isolated hepatocytes exposed to carbon tetrachloride, *Journal of Biological Chemistry*, 251, 4973, 1976.

327. **Lepay, D. A., Steinman, R. M., Nathan, C. F., Murray, H. W., and Cohn, Z. A.**, Liver macrophages in murine listeriosis. Cell-mediated immunity is correlated with an influx of macrophages capable of generating reactive oxygen intermediates, *Journal of Experimental Medicine*, 161, 1503, 1985.

328. **Billiar, T. R., Curran, R. D., Ferrari, F. K., Williams, D. L., and Simmons, R. L.**, Kupffer cell:hepatocyte coculture release nitric oxide in response to bacterial endotoxin, *Journal of Surgical Research*, 48, 349, 1990.

329. **Borchman, D., Paterson, C., and Delamere, N.**, Selective inhibition of membrane ATPase by hydrogen peroxide in the lens of the eye, in *Oxygen Radicals in Biology and Medicine*, Simic, M. G., Taylor, K. A., Ward, J. F., and von Sonntag, C., Eds., Plenum Press, New York, 1988, 1029.

330. **Imlay, J. A., and Linn, S.**, DNA damage and oxygen radical toxicity, *Science*, 240, 1302, 1988.

331. **Howard-Flonders, P.**, Inducible repair of DNA, *Scientific American*, 245, 72, 1981.

332. **Metzler, D. E.**, *Biochemistry: The Chemical Reactions of Living Cells*, Academic Press, New York, 1977, chap. 15.

333. **Gutteridge, J. M. C., and Halliwell, B.,** The antioxidant proteins of extracellular fluids, in *Cellular Antioxidant Defense Mechanisms,* Vol. II, Chow, C. K., Ed., CRC Press, Boca Raton, FL, 1988, chap. 15.

334. **Harris, E. D.,** Regulation of antioxidant enzymes, *FASEB Journal,* 6, 2675, 1992.

335. **Marx, J.,** Gene linked to Lou Gehrig's disease, *Science,* 259, 1393, 1993.

336. **Grisham, M. B., Perez, V. J., and Everse, J.,** Neuromelanogenic and cytotoxic properties of canine brainstem peroxidase, *Journal of Neurochemistry,* 48, 876, 1987.

337. **Okun, M. R., Edelstein, L. M., Or, N., Hamada, G., and Donnellan, B.,** The role of peroxidase vs. the role of tyrosinase in enzymatic conversion of tyrosine to melanin in melanocytes, mast cells and eosinophils, *Journal of Investigative Dermatology,* 55, 1, 1970.

338. **Okun, M., Edelstein, L., Or, N., Hamada, G., and Donnellan, B.,** Histochemical studies of conversion of tyrosine and DOPA to melanin mediated by mammalian peroxidase, *Life Science,* 9, 491, 1970.

339. **Rotilio, G., Paci, M., Sette, M., and Ciriolo, M. R.,** GSH as an antioxidant: new functions in the physiological activation of Cu, Zn superoxide dismutase, copper transport and transmembrane transduction of reducing power, in *Oxidative Damage & Repair. Chemical, Biological and Medical Aspects,* Davies, K. J. A., Ed., Pergamon Press, Oxford, 1991, 71.

340. **Pennington, J.,** Immune Activity and Immobilized Peroxidation in Administration of Novikoff Hepatoma Model. M.S. Thesis. Texas Tech University Health Sciences Center, Lubbock, 1982.

341. **Pileni, M.-P., and Santus, R.,** On the photosensitizing properties of *N*-formylkynurenine and related compounds, *Photochemistry and Photobiology,* 28, 525, 1978.

342. **Slawinska, D., and Slawinski, J.,** Low-level luminescence from biological objects, in *Chemi- and Bioluminescence,* Burr, J. G., Ed., Marcel Dekker, New York, 1985, chap. 13.

343. **Klausner, R. D., and Harford, J. B.,** Cis-trans models for post-transcriptional gene regulation, *Science,* 246, 870, 1989.

344. **Ananthan, J., Goldberg, A. L., and Voellmy, R.,** Abnormal proteins serve as eukaryotic stress signals and trigger the activation of heat shock genes, *Science,* 232, 522, 1986.

345. **Barbe, M. F., Tytell, M., Gower, D. J., and Welch, W. J.,** Hyperthermia protects against light damage in the rat retina, *Science,* 241, 1817, 1988.

346. **Lassmann, G., Liermann, B., and Langen, P.,** Stability and reactivation of tyrosine radicals from ribonucleotide reductase in tumor cells studied by ESR, *Free Radical Biology & Medicine,* 6, 9, 1989.

347. **Valerie, K., Delers, A., Bruck, C., Thiriatt, C., Rosenberg, H., Debouck, C., and Rosenberg, M.,** Activation of human immunodeficiency virus type I by DNA damage in human cells, *Nature,* 333, 78, 1988.

348. **Wallace, B. M., and Lasker, J. S.,** Awakenings ... UV light and HIV gene activation, *Science,* 257, 1211, 1992.

349. **Devary, Y., Rosette, C., Di Donato, J. A., and Karin, M.,** NF-kappa B activation by ultraviolet light not dependent on a nuclear signal, *Science,* 261, 1442, 1993.

350. **Voevodskaya, N. V., and Vanin, A. F.,** Gamma-irradiation potentiates L-arginine-dependent nitric oxide formation in mice, *Biochemical and Biophysical Research Communications,* 186, 1423, 1992.

351. **Blystone, R., Kiel, J., Parker, J., and Collumb, C.,** Expression of viral particles by RAW 264.7 mouse macrophage cells after treatment with lipopolysaccharide, in *Proceedings of the XIIth International Congress for Electron Microscopy,* San Francisco Press, San Francisco, 1990, 574.

352. **Stein, B., Kramer, M., Rahnsdorf, H. J., Ponta, H., and Henlich, P.,** UV-induced transcription from the human immunodeficiency virus type I (HIV-I) long terminal repeat and UV-induced secretion of an extracellular factor that induces HIV-I transcription in nonirradiated cells, *Journal of Virology,* 63, 4540, 1989.

353. **Ballard, D. W., Bohnlein, E., Lowenthal, J. W., Wano, Y., Franza, B. R., and Greene, W. C.,** HTLV-I tax induces cellular proteins that activate the kappa B element in the IL-2 receptor alpa gene, *Science,* 241, 1652, 1988.

354. **Kang, S.-M., Tran, A.-C., Grilli, M., and Lenardo, M. J.,** NF-kappa B subunit regulation in nontransformed CD4⁺ T lymphocytes, *Science,* 256, 1452, 1992.

355. **Holwitt, E. A.,** unpublished data, 1993.

356. **Brennan, J. K., Abboud, C. N., Di Persio, J. F., Barlow, G. H., and Lichtman, M. A.,** Autostimulation of growth by human myelogenous leukemia cells (HL-60), *Blood,* 58, 803, 1981.

357. **Ensoli, B., Nakamura, S., Salahuddin, S. Z., Biberfeld, P., Larsson, L., Beaver, B., Wong-Staal, F., and Gallo, R. C.,** AIDS-Kaposi's sarcoma-derived cells express cytokines with autocrine and paracrine growth effects, *Science,* 243, 223, 1989.

358. **Folks, T. M., Justement, J., Kinter, A., Dinarello, C. A., and Fauci, A. S.,** Cytokine-induced expression of HIV-I in a chronically infected promonocyte cell line, *Science,* 238, 800, 1987.

359. **Greene, W. C., Leonard, W. J., Wano, Y., Svetlik, P. B., Peffer, N. J., Sodroski, J. G., Rosen, C. A., Goh, W. C., and Haseltine, W. A.,** Trans-activator gene of HTLV-II induced IL-2 receptor and IL-2 cellular gene expression, *Science,* 232, 877, 1986.

360. **Ruben, S., Poteat, H., Tan, T.-H., Kawakami, K., Roeder, R., Haseltine, W., and Rosen, C. A.,** Cellular transcription factors and regulation of IL-2 receptor gene expression by HTLV-I tax gene product, *Science,* 241, 89, 1988.

361. **Schnittman, S. M., Lane, H. C., Higgins, S. E., Folks, T., and Fauci, A. S.,** Direct polyclonal activation of human B lymphocytes by the acquired immune deficiency syndrome virus, *Science,* 233, 1084, 1986.

362. **Boldogh, I., Abu Bakar, S., and Albrecht, T.,** Activation of proto-oncogenes: an immediate early event in human cytomegalovirus infection, *Science,* 247, 561, 1990.

363. **Koga, T., Wand-Wurttenberger, A., De Bruyn, J., Munk, M. E., Schoel, B., and Kaufmann, S. H. E.,** T cells against a bacterial heat shock protein recognize stressed macrophages, *Science,* 245, 1112, 1989.

364. **Akira, S., Hirano, T., Toga, T., and Kishimoto, T.,** Biology of multifunctional cytokines: IL-6 and related molecules (IL-1 and TNF), *FASEB Journal,* 4, 2860, 1990.

365. **Dinarello, C. A.,** Biology of interleukin 1, *FASEB Journal,* 2, 108, 1988.

366. **Warren, D. J., Slordal, L., and Moore, M. A. S.,** Tumor-necrosis factor induces cell cycle arrest in multipotential hematopoietic stem cells: a possible radioprotective mechanism, *European Journal of Hematology,* 45, 158, 1990.

367. **Neta, R., Oppenheim, J. J., Schreiber, R. D., Chizzonite, R., Ledney, G. D., and MacVittie, T. J.,** Role of cytokines (interleukin 1, tumor necrosis factor, and transforming growth factor beta) in natural and lipopolysaccharide-enhanced radioresistance, *Journal of Experimental Medicine,* 174, 1177, 1991.

368. **Neta, R., and Oppenheim, J. J.,** Anti-cytokine antibodies reveal the interdependence of pro-inflammatory cytokines in protection from lethal irradiation, in *New Advances on Cytokines,* Romognani, S., Mosmann, T. R., and Abbas, A. K., Eds., Raven Press, New York, 1992, 285.

369. **Sklar, M. D.,** The ras oncogenes increase the intrinsic resistance of NIH3T3 cells to ionizing radiation, *Science,* 239, 645, 1988.

370. **Besedovsky, H., Del Rey, A., Sorkin, E., and Dinarello, C. A.,** Immunoregulatory feedback between interleukin-1 and glucocorticoid hormones, *Science,* 233, 652, 1986.

371. **Bernton, E. W., Beach, J. E., Holaday, J. W., Smallridge, R. C., and Fein, H. G.,** Release of multiple hormones by a direct action of interleukin-1 on pituitary cells, *Science,* 238, 519, 1987.

372. **Berkenbosch, F., van Oers, J., Del Rey, A., Tilders, F., and Besedovsky, H.,** Corticotropin-releasing factor-producing neurons in the rat activated by interleukin-1, *Science,* 238, 524, 1987.

373. **Sopolsky, R., Rivier, C., Yamamoto, G., Plotsky, P., and Vale, W.,** Interleukin-1 stimulates secretion of hypothalamic corticotropin-releasing factor, *Science,* 238, 522, 1987.

374. **Breder, C. D., Dinarello, C. A., and Saper, C. B.,** Interleukin-1 immunoreactive innervation of the human hypothalamus, *Science,* 240, 321, 1988.

375. **Weber, R. J., and Pert, A.,** The periaqueductal gray matter mediates opiate-induced immunosuppression, *Science,* 245, 188, 1989.

376. **Su, T.-P., London, E. D., and Jaffe, J. H.,** Steroid binding at sigma receptors suggests a link between endocrine, nervous, and immune systems, *Science,* 240, 219, 1988.

377. **O'Dorisio, M. S.,** Biochemical characteristics of receptors for vasoactive intestinal polypeptide in nervous, endocrine, and immune systems, *Federation Proceedings,* 46, 192, 1987.

378. **Hiltz, M. E., and Lipton, J. M.,** Antiinflammatory activity of a COOH-terminal fragment of the neuropeptide alpha-MSH, *FASEB Journal,* 3, 2282, 1989.

379. **Gerschenfeld, H. K., and Weissman, I. L.,** Cloning of a cDNA for a T cell-specific serine protease from a cytotoxic T lymphocyte, *Science,* 232, 854, 1986.

380. **Edelman, A. S., and Zolla-Pazner, S.,** AIDS: a syndrome of immune dysregulation, dysfunction, and deficiency, *FASEB Journal,* 3, 22, 1989.

381. **Koenig, S., Gendelman, H. E., Orenstein, J. M., Dal Canto, M. C., Pezeshkpour, G. H., Yungbluth, M., Janotta, F., Aksamit, A., Martin, M. A., and Fauci, A. S.,** Detection of AIDS virus in macrophages in brain tissue from AIDS patients with encephalopathy, *Science,* 233, 1089, 1986.

382. **Li, J. J., and Li, S. A.,** Estrogen carcinogenesis in Syrian hamster tissues: role of metabolism, *Federation Proceedings,* 46, 1858, 1987.

383. **Horowitz, M. C.,** Cytokines and estrogen in bone: anti-osteoporotic effects, *Science,* 260, 626, 1993.

384. **Kiel, J. L.,** Paget's disease in snakes, in *Annual Proceedings of the American Association of Zoo Veterinarians,* Fowler, M. E., Ed., AAZV, Tampa, FL, 1984, 201.

385. **Jilka, R. L., Hangoc, G., Girasole, G., Passeri, G., Williams, D. C., Abrams, J. S., Boyce, B., Broxmeyer, H., and Manolagas, S. C.,** Increased osteoclast development after estrogen loss: mediation by interleukin-6, *Science,* 257, 88, 1992.

386. **Rier, S. E., Martin, D. C., Bowman, R. E., Dmowski, W. P., and Becker, J. L.,** Endometriosis in rhesus monkeys (*Macaca mulatta*) following chronic exposure to 2,3,7,8-tetrachlorodibenzo-p-dioxin, *Fundamental and Applied Toxicology,* 21, 433, 1993.

387. **Rier, S., Spangelo. B., Martin, D., Bowman, R., and Becker, J.,** Tumor necrosis factor alpha and interleukin-6 production by peripheral blood mononuclear cells from rhesus monkeys with endometriosis, presented at the American Association of Immunology/Cellular Immunology Society Annual Meeting, 1993.

388. **Keller, R., Keist, R., and Hecker, E.,** Tumour-promoting diterpene esters prevent macrophage activation, *British Journal of Cancer,* 45, 144, 1982.

389. **Goetzl, E. J.,** The conversion of leukotriene C_4 to isomers of leukotriene B_4 by human eosinophil peroxidase, *Biochemical and Biophysical Research Communications,* 106, 270, 1982.

390. **Miyazawa, K., Kiyono, S., and Inoue, K.,** Modulation of stimulus-dependent human platelet activation by C-reactive protein modified with active oxygen species, *Journal of Immunology,* 141, 570, 1988.

391. **Labib, R. S., and Tomasi, Jr., T. B.,** Enzymatic oxidation of polyamines. Relationship to immunosuppressive properties, *European Journal of Immunology,* 11, 266, 1981.

392. **Britigan, B. E., and Hamill, D. R.,** Effect of the spin trap 5,5 dimethyl-1-pyrroline-*N*-oxide (DMPO) on human neutrophil function: novel inhibition of neutrophil stimulus-response coupling? *Free Radical Biology & Medicine,* 8, 459, 1990.

393. **Chen, Y.-N. P., and Marnett, L. J.,** Heme prosthetic group required for acetylation of prostaglandin H synthase by aspirin, *FASEB Journal,* 3, 2294, 1989.

394. **Justement, L. B., Aldrich, W. A., Wenger, G. D., O'Dorisio, M. S., and Zwilling, B. S.,** Modulation of cyclic AMP-dependent protein kinase isozyme expression associated with activation of a macrophage cell line, *Journal of Immunology,* 136, 270, 1986.

395. **Vallera, D. A., Gamble, C. E., and Schmidtke, J. R.,** Lipopolysaccharide-induced immunomodulation of the generation of cell-mediated cytotoxicity. II. Evidence for the involvement of a regulatory B lymphocyte, *Journal of Immunology,* 124, 641, 1980.

396. **Kunkel, S. L., Chensue, S. W., and Phan, S. H.,** Prostaglandins as endogenous mediators of interleukin-1 production, *Journal of Immunology,* 136, 186, 1986.

397. **Pruett, S. B.,** Model Systems for Assessing the Effects of Microwave Radiation on the Immune System. Final Report on Contract No. F49620–85-C-0013/SB5851–0360, Office of Scientific Research, U.S. Air Force, Bolling AFB, Washington, DC, 1988.

398. **Kazlauskas, A., Ellis, C., Pawson, T., and Cooper, J. A.,** Binding of GAP to activated PDGF receptors, *Science,* 247, 1578, 1990.

399. **Raines, E. W., Dower, S. K., and Ross, R.,** Interleukin-1 mitogenic activity for fibroblasts and smooth muscle cells is due to PDGF-AA, *Science,* 243, 393, 1989.

400. **Schindler, C., Shuai, K., Prezioso, V. R., and Darnell, J. E., Jr.,** Interferon-dependent tyrosine phosphorylation of a latent cytoplasmic transcription factor, *Science,* 257, 809, 1992.

401. **Marx, J.,** Taking a direct path to the genes, *Science,* 257, 744, 1992.

402. **Feig, L. A.,** The many roads that lead to ras, *Science,* 260, 767, 1993.

403. **Rozengurt, E.,** Early signals in the mitogenic response, *Science,* 234, 161, 1986.

404. **Levi-Montalcini, R., and Calissano, P.,** The nerve growth factor, *Scientific American,* 240, 68, 1979.

405. **Kaplan, D. R., Hempstead, B. L., Martin-Zanca, D., Chao, M. V., and Parada, L. F.,** The trk proto-oncogene product: a signal transducing receptor for nerve growth factor, *Science,* 252, 554, 1991.

406. **Kavanagh, T. J., Grossmann, A., Jinneman, J. C., Kanner, S. B., White, C. C., Eaton, D. L., Ledbetter, J. A., and Rabinovitch, P. S.,** The effect of 1-chloro-2,4-dinitrobenzene exposure on antigen receptor (CD3)-stimulated transmembrane signal transduction in purified subsets of human peripheral blood lymphocytes, *Toxicology and Applied Pharmacology,* 119, 91, 1993.

407. **Coffey, R. G., and Hadden, J. W.,** Stimulation of lymphocyte guanylate cyclase by arachidonic acid and HETES, in *Prostaglandins, Leukotrienes, and Lipoxins: Biochemistry, Mechanism of Action and Clinical Applications,* Bailey, J. M., Ed., Plenum Press, New York, 1985, chap. 48.

408. **De Rubertis, F. R., and Craven, P. A.,** Activation of guanylate cyclase by *N*-nitroso carcinogens and related agonists: evidence for a free radical mechanism, in *Free Radicals and Cancer,* Floyd, R. A., Ed., Marcel Dekker, New York, 1982, chap. 7.

409. **Gougeon, M.-L., Garcia, S., Heeney, J., Tachopp, R., Lecoeur, H., Guetard, D., Rame, V., Dauguet, C., and Montagnier, L.,** Programmed cell death in AIDS-related HIV and SIV infections, *AIDS Research and Human Retroviruses,* 9, 553, 1993.

410. **van der Brugge-Gamelkoorn, G. J., van de Ende, M., and Sminia, T.,** Changes occurring in the epithelium covering the bronchus-associated lymphoid tissue of rats after intratracheal challenge with horseradish peroxidase, *Cell and Tissue Research,* 245, 439, 1986.

411. **Fairbairn, L. J., Cowling, G. J., Reipert, B. M., and Dexter, T. M.,** Suppression of apoptosis allows differentiation and development of a multipotential hemopoietic cell line in the absence of added growth factors, *Cell,* 74, 823, 1993.

412. **Ding, A., Nathan, C. F., Graycar, J., Derynck, R., Stuehr, D. J., and Srimal, S.,** Macrophage deactivating factor and transforming growth factors-beta 1, -beta 2, and -beta 3, inhibit induction of macrophage nitrogen oxide synthesis by IFN-gamma, *Journal of Immunology,* 145, 940, 1990.

413. **Rappolee, D. A., Mark, D., Bonda, M. J., and Werb, Z.,** Wound macrophages express TGF-alpha and other growth factors in vivo: analysis by mRNA phenotyping, *Science,* 241, 708, 1988.

414. **Baulieu, E.-E.,** Contragestion and other clinical applications of RU486, an antiprogesterone at the receptor, *Science,* 245, 1351, 1989.

415. **Kaczmarek, L., Calabretta, B., Kao, H.-T., Heintz, N., Nevins, J., and Baserga, R.,** Control of hsp70 RNA levels in human lymphocytes, *Journal of Cell Biology,* 104, 183, 1987.

416. **Wu, B. J., and Morimoto, R. I.,** Transcription of the human hsp70 gene is induced by serum stimulation, *Proceedings of the National Academy of Science of the USA,* 82, 6070, 1985.

417. **Wu, B. J., Kingston, R. E., and Morimoto, R. I.,** Human HSP70 promoter contains at least two distinct regulatory domains, *Proceedings of the National Academy of Sciences of the USA,* 83, 629, 1986.

418. **Ratcliffe, N. A., Leonard, C., and Rowley, A. F.,** Prophenoloxidase activation: nonself recognition and cell cooperation in insect immunity, *Science,* 226, 557, 1984.

419. **Bruno, J. G., and Kiel, J. L.,** Synthesis of diazoluminomelanin (DALM) in HL-60 cells for possible use as a cellular-level microwave dosimeter, *Bioelectromagnetics,* 15, 315, 1994.

420. **Wright, C. G., and Lee, D. H.,** Pigmented epithelial cells of the membranous saccular wall of the chinchilla, *Acta Otolaryngologica,* 102, 438, 1986.

421. **La Ferriere, K. A., Arenberg, I. K., Hawkins, J. E., Jr., and Johnsson, L.-G.,** Melanocytes of the vestibular labyrinth and their relationship to the microvasculature, *Annals of Otology, Rhinology and Laryngology,* 83, 685, 1974.

422. **Hilding, D. A., and Ginzberg, R. D.,** Pigmentation of the stria vascularis. The contribution of the neural crest melanocytes, *Acta Otolaryngologica,* 84, 24, 1977.

423. **d'Ischia, M., Napolitano, A., and Prota, G.,** Peroxidase as an alternative to tyrosinase in the oxidative polymerization of 5,6-dihydroxyindoles to melanin(s), *Biochimica et Biophysica Acta,* 1073, 423, 1991.

424. **Hamilton, G. A., and Libby, R. D.,** The valence of copper and the role of superoxide in the D-galactose oxidase catalyzed reaction, *Biochemical and Biophysical Research Communications,* 55, 333, 1973.

425. **Metzler, D. E.,** *Biochemistry: The Chemical Reactions of Living Cells,* Academic Press, New York, 1977, chap. 8.

426. **Carter, S. K., Bakowski, M. T., and Hellmann, K.,** *Chemotherapy of Cancer,* 2nd Ed., John Wiley & Sons, New York, 1981, 182.

427. **Korthuis, R. J., and Granger, D. N.,** Ischemia-reperfusion injury: role of oxygen-derived free radicals, in Physiology of *Oxygen Radicals,* Taylor, A. E., Matalon, S., and Ward, P., Eds., American Physiological Society, Bethesda, MD, 1986, chap. 17.

428. **Granger, D. N., and Parks, D. A.,** Role of oxygen radicals in the pathogenesis of intestinal ischemia, *The Physiologist,* 26, 159, 1983.

429. **Devenyi, Z. J., Orchard, J. L., and Powers, R. E.,** Xanthine oxidase in mouse pancreas: effects of caerulein-induced acute pancreatitis, *Biochemical and Biophysical Research Communications,* 149, 841, 1987.

430. **Engler, R.,** Granulocytes and oxidative injury in myocardial ischemia and reperfusion. Introduction, *Federation Proceedings,* 46, 2395, 1987.

431. **Mullane, K. M., Salmon, J. A., and Kraemer, R.,** Leukocyte-derived metabolites of arachidonic acid in ischemia-induced myocardial injury, *Federation Proceedings,* 46, 2422, 1987.

432. **Ward, P. A., Johnson, K. J., and Till, G. O.**, Oxygen radicals, neutrophils, and acute tissue injury, in *Physiology of Oxygen Radicals*, Taylor, A. E., Matalon, S., and Ward, P., Eds., American Physiological Society, Bethesda, MD, 1986, chap. 10.

433. **Taylor, A. E., and Martin, D.**, Oxygen radicals and the microcirculation, *The Physiologist*, 26, 152, 1983.

434. **McCord, J. M.**, The biochemistry and pathophysiology of superoxide, *The Physiologist*, 26, 156, 1983.

435. **Crapo, J. D., Freeman, B. A., Barry, B. E., Turrens, J. F., and Young, S. L.**, Mechanisms of hyperoxic injury to the pulmonary microcirculation, *The Physiologist*, 26, 170, 1983.

436. **Repine, J. E., and Tate, R. M.**, Oxygen radicals and lung edema, *The Physiologist*, 26, 177, 1983.

437. **Lin, H., McFaul, S. J., Brady, J. C., and Everse, J.**, The mechanism of peroxidase-mediated cytotoxicity. II. Role of the heme moiety, *Proceedings of the Society for Experimental Biology and Medicine*, 187, 7, 1988.

438. **Murray, H. W., Juangbhanich, C. W., Nathan, C. F., and Cohn, Z. A.**, Macrophage oxygen-dependent antimicrobial activity. II. The role of oxygen intermediates, *Journal of Experimental Medicine*, 150, 950, 1979.

439. **Murray, H. W., and Cohn, Z. A.**, Macrophage oxygen-dependent antimicrobial activity. III. Enhanced oxidative metabolism as an expression of macrophage activation, *Journal of Experimental Medicine*, 152, 1596, 1980.

440. **Nathan, C. F.**, Mechanisms of macrophage antimicrobial activity, *Transactions of the Royal Society of Tropical Medicine and Hygiene*, 77, 620, 1983.

441. **Grisham, M. B., and McCord, J. M.**, Chemistry and cytotoxicity of reactive oxygen metabolites, in *Physiology of Oxygen Radicals*, Taylor, A. E., Matalon, S., and Ward, P., Eds., American Physiological Society, Bethesda, MD, 1986, chap. 1.

442. **Marx, R. E., Kline, S. N., Ehler, W. G., and Kiel, J. L.**, Induced regression of established oral-carcinoma and suppression of dysplastic mucosal changes by oxygen radical generating systems: a blinded study of glucose oxidase-horseradish peroxidase and hyperbaric oxygen, *Journal of Oral and Maxillofacial Surgery*, 44, M6, 1986.

443. **Ehler, W. J., Marx, R. E., Kiel, J., Ravelo, J. I., Kline, S. N., and Cissik, J. H.**, Induced regression of oral carcinoma by oxygen radical generating systems, *Journal of Hyperbaric Medicine*, 6, 111, 1991.

444. **Everse, K. E., Lin, H., Stuyt, E. L., Brady, J. C., Buddingh, F., and Everse, J.**, Antitumor activity of peroxidase, British *Journal of Cancer*, 51, 743, 1985.

445. **Bigley, N. J., Curiel, R. E., Kozlowski, D., Ishikawa, R., Henderson, R. A., and Kiel, J. L.**, Insoluble crosslinked cytotoxic oxidase-peroxidase system (ICCOPS) and spleen cell-picornavirus interactions, presented at the 75th Annual Meeting of the Federation of American Societies for Experimental Biology, Atlanta, April 21–25, 1991.

446. **Wei, R.-Q., Lefkowitz, S. S., Lefkowitz, D. L., and Everse, J.**, Activation of macrophages by peroxidases, *Proceedings of the Society for Experimental Biology and Medicine*, 182, 515, 1986.

447. **Lefkowitz, D. L., Lefkowitz, S. S., Mone, J., and Everse, J.**, Peroxidase-induced enhancement of chemiluminescence by murine peritoneal macrophages, *Life Sciences*, 43, 739, 1988.

448. **Lin, H., and Everse, J.**, The cytotoxic activity of hematoporphyrin: studies on the possible role of transition metals, *Biochemical Medicine and Metabolic Biology*, 36, 60, 1986.

449. **Snyder, S. H., and Bredt, D. S.**, Biological roles of nitric oxide, *Scientific American*, 266, 68, 1992.

450. **Alper, J.**, NO role in immune responses being recognized, *ASM News*, 59, 9, 1993.

451. **Kwon, N. S., Nathan, C. F., Gilker, C., Griffith, O. W., Matthews, D. E., and Stuehr, D. J.,** L-Citrulline production from L-arginine by macrophage nitric oxide synthase. The ureido oxygen derives from dioxygen, *Journal of Biological Chemistry,* 265, 13442, 1990.

452. **Stuehr, D. J., Kwon, N. S., and Nathan, C. F.,** FAD and GSH participate in macrophage synthesis of nitric oxide, *Biochemical and Biophysical Research Communications,* 168, 558, 1990.

453. **White, K. A., and Marletta, M. A.,** Nitric oxide synthase is a cytochrome P-450 type hemoprotein, *Biochemistry,* 31, 6627, 1992.

454. **Lancaster, Jr., J. R., and Hibbs, Jr., J. B.,** EPR demonstration of iron-nitrosyl complex formation by cytotoxic activated macrophages, *Proceedings of the National Academy of Sciences of the USA,* 87, 1223, 1990.

455. **Drapier, J.-C., Pellat, C., and Henry, Y.,** Generation of EPR-detectable nitrosyl-iron complexes in tumor target cells cocultured with activated macrophages, *Journal of Biological Chemistry,* 266, 10162, 1991.

456. **Amber, I. J., Hibbs, J. B., Jr., Taintor, R. R., and Vavrin, Z.,** The l-arginine dependent effector mechanism is induced in murine adenocarcinoma cells by culture supernatant from cytotoxic activated macrophages, *Journal of Leukocyte Biology,* 43, 187, 1988.

457. **Barbul, A.,** Physiology and pharmacology of arginine, in *Nitric Oxide From L-Arginine: A Bioregulatory System,* Moncada, S., and Higgs, E. A., Eds., Elsevier, Amsterdam, 1990, chap. 33.

458. **Kolb, H., and Kolb-Bachofen, V.,** Nitric oxide: a pathogenic factor in autoimmunity, *Immunology Today,* 13, 157, 1992.

459. **Kwon, N. S., Stuehr, D. J., and Nathan, C. F.,** Inactivation of ribonucleotide reductase (RR) by macrophage nitric oxide (NO·) synthase and generation of NO· by the RR inhibitor hydroxyurea, *FASEB Journal,* 5, A1346, Abstr. 5655, 1991.

460. **Drapier, J.-C., and Hibbs, J. B., Jr.,** Differentiation of murine macrophages to express nonspecific cytotoxicity for tumor cells results in L-arginine-dependent inhibition of mitochondrial iron-sulfur enzymes in the macrophage effector cells, *Journal of Immunology,* 140, 2829, 1988.

461. **Ding, A. H., Nathan, C. F., and Stuehr, D. J.,** Release of reactive nitrogen intermediates and reactive oxygen intermediates from mouse peritonealmacrophages. Comparison of activating cytokines and evidence for independent production, *Journal of Immunology,* 141, 2407, 1988.

462. **Xie, Q.-W., Cho, H. J., Calaycay, J., Mumford, R. A., Swiderek, K. M., Lee, T. D., Ding, A., Troso, T., and Nathan, C.,** Cloning and characterization of inducible nitric oxide synthase from mouse macrophages, *Science,* 256, 225, 1992.

463. **Bredt, D. S., Ferris, C. D., and Snyder, S. H.,** Nitric oxide synthase regulatory sites. Phosphorylation by cyclic AMP-dependent protein kinase, protein kinase C, and calcium/calmodulin protein kinase: identification of flavin and calmodulin binding sites, *Journal of Biological Chemistry,* 267, 10976, 1992.

464. **Lyons, C. R., Orloff, G. J., and Cunningham, J. M.,** Molecular cloning and functional expression of an inducible nitric oxide synthase from a murine macrophage cell line, *Journal of Biological Chemistry,* 267, 6370, 1992.

465. **Bredt, D. S., Hwang, P. M., Glatt, C. E., Lowenstein, C., Reed R. R., and Snyder, S. H.,** Cloned and expressed nitric oxide synthase structurally resembles cytochrome P-450 reductase, *Nature,* 351, 714, 1991.

466. **McMillan, K., Bredt, D. S., Hirsch, D. J., Snyder, S. H., Clark, J. E., and Masters, B. S. S.,** Cloned, expressed rat cerebellar nitric oxide synthase contains stoichiometric amounts of heme, which binds carbon monoxide, *Proceedings of the National Academy of Sciences of the USA,* 89, 11141, 1992.

467. Stuehr, D. J., Cho, H. J., Kwon, N. S., Weise, M. F., and Nathan, C. F., Purification and characterization of the cytokine-induced macrophage nitric oxide synthase: an FAD- and FMN-containing flavoprotein, *Proceedings of the National Academy of Sciences of the USA,* 88, 7773, 1991.

468. Kwon, N. S., Nathan, C. F., and Stuehr, D. J., Reduced biopterin as a cofactor in the generation of nitrogen oxides by murine macrophages, *Journal of Biological Chemistry,* 264, 20496, 1989.

469. Dimmeler, S., and Brune, B., L-Arginine stimulates an endogenous ADP-ribosyl-transferase, *Biochemical and Biophysical Research Communications,* 178, 848, 1991.

470. Schmidt, H. H. H. W., Warner, T. D., Nakane, M., Forstermann, U., and Murad, F., Regulation and subcellular location of nitrogen oxide synthases in RAW 264.7 macrophages, *Molecular Pharmacology,* 41, 615, 1992.

471. Moncada, S., Palmer, R. M. J., and Higgs, E. A., The biological significance of nitric oxide formation from L-arginine, *Biochemical Society Transactions,* 17, 642, 1989.

472. Wink, D. A., Kasprzak, K. S., Maragos, C. M., Elespuru, R. K., Misra, M., Dunams, T. M., Cebula, T. A., Koch, W. H., Andrews, A. W., Allen, J. S., and Keefer, L. K., DNA deaminating ability and genotoxicity of nitric oxide and its progenitors, *Science,* 254, 1001, 1991.

473. Stamler, J. S., Jaraki, O., Osborne, J., Simon, D. I., Keaney, J., Vita, J., Singel, D., Valeri, C. R., and Loscalzo, J., Nitric oxide circulates in mammalian plasma primarily as an *S*-nitroso adduct of serum albumin, *Proceedings of the National Academy of Sciences of the USA,* 89, 7674, 1992.

474. Vanin, A. F., Endothelium-derived relaxing factor is a nitrosyl iron complex with thiol ligands, *FEBS Letters,* 289, 1, 1991.

475. Holmgren, A., Thioredoxins: structure and functions, *Trends in Biological Science,* January, 26, 1981.

476. Buchanan, B. B., Wolosiuk, R. A., and Schurmann, P., Thioredoxin and enzyme regulation, *Trends in Biological Science,* April, 93, 1979.

477. Brune, B., and Lapetina, E. G., Activation of a cytosolic ADP-ribosyltransferase by nitric oxide-generating agents, *Journal of Biological Chemistry,* 264, 8455, 1989.

478. Okayama, H., Edson, C. M., Fukushima, M., Ueda, K., and Hayaisha, O., Purification and properties of poly(adenosine diphosphate ribose) synthetase, *Journal of Biological Chemistry,* 252, 7000, 1977.

479. Zhang, X., and Morrison, D. C., Pertussis toxin-sensitive factor differentially regulates lipopolysaccharide-induced tumor necrosis factor-alpha and nitric oxide production in mouse peritoneal macrophages, *Journal of Immunology,* 150, 1011, 1993.

480. Keller, R., Keist, R., Klauser, S., and Schweiger, A., The macrophage response to bacteria: flow of L-arginine through the nitric oxide and urea pathways and induction of tumoricidal activity, *Biochemical and Biophysical Research Communications,* 177, 821, 1991.

481. Hibbs, Jr., J. B., Taintor, R. R., Vavrin, Z., and Rachlin, E. M., Nitric oxide: a cytotoxic activated macrophage effector molecule, *Biochemical and Biophysical Research Communications,* 157, 87, 1988.

482. Amber, I. J., Hibbs, Jr., J. B., Parker, C. J., Johnson, B. B., Taintor, R. R., and Vavrin, Z., Activated macrophage conditioned medium: identification of the soluble factors inducing cytotoxicity and the L-arginine dependent effector mechanism, *Journal of Leukocyte Biology,* 49, 610, 1991.

483. Liew, F. Y., Millott, S., Parkinson, C., Palmer, R. M. J., and Moncada, S., Macrophage killing of *Leishmania* parasite *in vivo* is mediated by nitric oxide from L-arginine, *Journal of Immunology,* 144, 4794, 1990.

484. Karupiah, G., Xie, Q.-W., Buller, R. M. L., Nathan, C., Duarte, C., and MacMicking, J. D., Inhibition of viral replication by interferon-gamma-induced nitric oxide synthase, *Science,* 261, 1445, 1993.

485. **Croen, K. D.,** Evidence for an antiviral effect of nitric oxide. Inhibition of herpes simplex virus type 1 replication, *Journal of Clinical Investigations,* 91, 2446, 1993.

486. **Schuman, E. M., and Madison, D. V.,** A requirement for the intercellular messenger nitric oxide in long-term potentiation, *Science,* 254, 1503, 1991.

487. **Barinaga, M.,** Is nitric oxide the "retrograde messenger"? *Science,* 254, 1296, 1991.

488. **Snyder, S. H.,** Nitric oxide: first in a new class of neurotransmitters? *Science,* 257, 494, 1992.

489. **Knowles, R. G., Palacios, M., Palmer, R. M. J., and Moncada, S.,** Nitric oxide synthase in the brain, in *Nitric Oxide From L-Arginine: A Bioregulatory System,* Moncada, S., and Higgs, E. A., Eds., Elsevier, Amsterdam, 1990, chap. 16.

490. **Moncada, S., Palmer, R. M. J., and Higgs, E. A.,** Biosynthesis of nitric oxide from L-arginine. A pathway for the regulation of cell function and communication, *Biochemical Pharmacology,* 38, 1709, 1989.

491. **Gally, J. A., Montague, P. R., Reeke, G. N., Jr., and Edelman, G. M.,** The NO hypothesis: possible effects of a short-lived, rapidly diffusable signal in the development and function of the nervous system, *Proceedings of the National Academy of Sciences, USA,* 87, 3547, 1990.

492. **Palmer, R. M. J., Ferrige, A. G., and Moncada, S.,** Nitric oxide release accounts for the biological activity of endothelium-derived relaxing factor, *Nature,* 327, 524, 1987.

493. **Baum, R.,** Complexes control nitric oxide release, *Chemical and Engineering News,* September 14, 32, 1992.

494. **Furchgott, R. F., Jothianandan, D., and Freoy, D.,** Endothelium-derived relaxing factor: some old and new findings, in *Nitric Oxide From L-Arginine: A Bioregulatory System,* Moncada, S., and Higgs, E. A., Eds., Elsevier, Amsterdam, 1990, chap. 2.

495. **Tare, M., Parkington, H. C., Coleman, H. A., Neild, T. O., and Dusting, G. J.,** Actions of endothelium-derived nitric oxide include hyperpolarization of vascular smooth muscle, in *Nitric Oxide From L-Arginine: A Bioregulatory System,* Moncada, S., and Higgs, E. A., Eds., Elsevier, Amsterdam, 1990, chap. 9.

496. **Kubes, P. Suzuki, M., and Granger, D. N.,** Nitric oxide: an endogenous modulator of leukocyte adhesion, *Proceedings of the National Academy of Sciences of the USA,* 88, 4651, 1991.

497. **Takasawa, S., Nata, k., Yonekura, H., and Okamoto, H.,** Cyclic ADP-ribose in insulin secretion from pancreatic beta cells, *Science,* 259, 370, 1993.

498. **Payne, C. M., Bjore, C. G., Jr., and Schultz, D. A.,** Change in the frequency of apoptosis after low- and high-dose X-irradiation of human lymphocytes, *Journal of Leukocyte Biology,* 52, 433, 1992.

499. **Shi, Y. Glynn, J. M., Guilbert, L. J., Cotter, T. G., Bissonnette, R. P., and Green, D. R.,** Role for c-myc in activation-induced apoptotic cell death in T cell hybridomas, *Science,* 257, 212, 1992.

500. **Raff, M. C., Barres, B. A., Burne, J. F., Coles, H. S., Ishizaki, Y., and Jacobson, M. D.,** Programmed cell death and the control of cell survival: lessons from the nervous system, *Science,* 262, 695, 1993.

501. **Mercep, M., Weissman, A. M., Frank, S. J., Klausner, R. D., and Ashwell, J. D.,** Activation-driven programmed cell death and T cell receptor zeta eta expression, *Science,* 246, 1162, 1989.

502. **Gougeon, M.-L., and Montagnier, L.,** Apoptosis in AIDS, *Science,* 260, 1269, 1993.

503. **Meyaard, L., Otto, S. A., Jonker, R. R., Mijnster, M. J., Keet, R. P. M., and Miedema, F.,** Programmed death of T cells in HIV-1 infection, *Science,* 257, 217, 1992.

504. **Galione, A.,** Cyclic ADP-ribose: a new way to control calcium, *Science,* 259, 325, 1993.

505. **Sun, S.-C., Ganchi, P. A., Ballard, D. W., and Greene, W. C.,** NF-kappa B controls expression of inhibitor I kappa B alpha: evidence for an inducible autoregulatory pathway, *Science,* 259, 1912, 1993.

506. **Milburn, M. V., Tong, L., de Vos, A. M., Brunger, A., Yamaizumi, Z., Nishimura, S., and Kim, S.-H.,** Molecular switch for signal transduction: structural differences between active and inactive forms of protooncogenic ras proteins, *Science,* 247, 939, 1990.

507. **van Voorhis, W. C., Witmer, M. D., and Steinman, R. M.,** The phenotype of dendritic cells and macrophages, *Federation Proceedings,* 42, 3114, 1983.

508. **Cameron, P. U., Freudenthal, P. S., Barker, J. M., Gezelter, S., Inaba, K., and Steinman, R. M.,** Dendritic cells exposed to human immunodeficiency virus type-1 transmit a vigorous cytopathic infection to CD4+ T cells, *Science,* 257, 383, 1992.

509. **Mosier, D. E., Gulizia, R. J., MacIsaac. P. D., Torbett, B. E., and Levy, J. A.,** Rapid loss of CD4+ T cells in human-PBL-SCID mice by noncytopathic HIV isolates, *Science,* 260, 689, 1993.

510. **Schmidt, H. H. H. W., Seifert, R., and Bohme, E.,** Formation and release of nitric oxide from human neutrophils and HL-60 cells induced by a chemotactic peptide, platelet activating factor and leukotriene B_4, *FEBS Letters,* 244, 357, 1989.

511. **Fleming, S. D., Iandolo, J. J., and Chopes, S. K.,** Murine macrophage activation by staphylococcal exotoxins, *Infection and Immunity,* 59, 4049, 1991.

512. **Moncada, S., and Higgs, A.,** The L-arginine-nitric oxide pathway, *The New England Journal of Medicine,* 329, 2002, 1993.

513. **Marletta, M. A., Yoon, P. S., Iyengar, R., Leaf, C. D., and Wishnok, J. S.,** Macrophage oxidation of L-arginine to nitrite and nitrate: nitric oxide is an intermediate, *Biochemistry,* 27, 8706, 1988.

514. **Smith, R. P., Louis, C. A., Kruszyna, R., and Kruszyna, H.,** Acute neurotoxicity of sodium azide and nitric oxide, *Fundamental and Applied Toxicology,* 17, 120, 1991.

515. **Klatt, P., Heinzel, B., John, M., Kastner, M., Bohme, E., and Mayer, B.,** Ca^{2+}/calmodulin-dependent cytochrome c reductase activity of brain nitric oxide synthase, *Journal of Biological Chemistry,* 267, 11374, 1992.

516. **Stuehr, D. J., Kwon, N. S., Gross, S. S., Thiel, B. A., Levi, R., and Nathan, C. F.,** Synthesis of nitrogen oxides from L-arginine by macrophage cytosol: requirement for inducible and constitutive components, *Biochemical and Biophysical Research Communications,* 161, 420, 1989.

517. **Trump, B. F.,** Programmed cell death and Ca^{2+} signaling, *Neuroscience Facts,* 3, 75, 1992.

518. **Hockenberg, D. M., Oltvai, Z. N., Yin, X.-M., Milliman, C. L., and Korsmeyer, S. J.,** Bcl-2 functions in an antioxidant pathway to prevent apoptosis, *Cell,* 75, 241, 1993.

519. **Veis, D. J., Sorenson, C. M., Shutter, J. R., and Korsmeyer, S. J.,** Bcl-2-deficient mice demonstrate fulminant lymphoid apoptosis, polycystic kidneys, and hypopigmented hair, *Cell,* 75, 229, 1993.

520. **Pruett, S. B., Obiri, N., and Kiel, J. L.,** Involvement and relative importance of at least two distinct mechanisms in the effect of 2-mercaptoethanol on murine lymphocytes in culture, *Journal of Cellular Physiology,* 141, 40, 1989.

521. **Bruno, J. G., Parker, J. E., and Kiel, J. L.,** Plant nitrate reductase gene fragments enhance nitrite production in activated murine macrophage cell lines, *Biochemical and Biophysical Research Communications,* 201, 284, 1994.

522. **Lander, H. M., Sehajpal, P., Levine, D. M., and Novogrodsky, A.,** Activation of human peripheral blood mononuclear cells by nitric oxide-generating compounds, *Journal of Immunology,* 150, 1509, 1993.

523. **Langrehr, J. M., Murase, N., Markus, P. M., Cai, X., Neuhaus, P., Schraut, W., Simmons, R. L., and Hoffman, R. A.,** Nitric oxide production in host-versus-graft and graft-versus-host reactions, *Journal of Clinical Investigations,* 90, 679, 1992.

524. **Gazzinelli, R. T., Oswald, I. P., James, S. L., and Sher, A.,** IL-10 inhibits parasite killing and nitrogen oxide production by IFN-gamma activated macrophages, *Journal of Immunology,* 148, 1792, 1992.

525. **Dixit, V. M., Marks, R. M., Sarma, V., and Prochownik, E. V.,** The antimitogenic action of tumor necrosis factor is associated with increased AP-1/c-jun proto-oncogene transcription, *Journal of Biological Chemistry,* 264, 16905, 1989.

526. **Albina, J. E., and Henry, Jr., W. L.,** Suppression of lymphocyte proliferation through the nitric oxide synthesizing pathway, *Journal of Surgical Research,* 50, 403, 1991.

527. **Shuai, K., Schindler, C., Prezioso, V. R., and Darnell, J. E.,Jr.,** Activation of transcription by IFN-gamma: tyrosine phosphorylation of a 91-kD DNA binding protein, *Science,* 258, 1808, 1992.

528. **Virca, G. D., Kim, S. Y., Glaser, K. B., and Ulevitch, R. J.,** Lipopolysaccharide induces hyporesponsiveness to its own action in RAW 264.7 cells, *Journal of Biological Chemistry,* 264, 21951, 1989.

529. **Minty, A., Chalon, P., Derocq, J.-M., Dumont, X., Guillemot, J.-C., Kaghad, M., Labit, C., Leplatois, P., Liauzun, P., Miloux, B., Minty, C., Casellas, P., Loison, G., Lupker, J., Shire, D., Ferrara, P., and Caput, D.,** Interleukin-13 is a new human lymphokine regulating inflammatory and immune responses, *Nature,* 362, 248, 1993.

530. **Freedman, M. H., and Raff, M. C.,** Induction of increased calcium uptake in mouse T lymphocytes by concanavalin A and its modulation by cyclic nucleotides, *Nature,* 255, 378, 1975.

531. **Li, L., Kilbourn, R. G., Adams, J., and Fidler, I. J.,** Role of nitric oxide in lysis of tumor cells by cytokine-activated endothelial cells, *Cancer Research,* 51, 2531, 1991.

532. **Ribeiro, J. M. C., Hazzard, J. M. H., Nussenzveig, R. H., Champagne, D. E., and Walker, F. A.,** Reversible binding of nitric oxide by a salivary heme protein from a bloodsucking insect, *Science,* 260, 539, 1993.

533. **Ignarro, l. J.,** Endothelium-derived nitric oxide: actions and properties, *FASEB Journal,* 3, 31, 1989.

534. **Burnett, A. L., Lowenstein, C. J., Bredt, D. S., Chang, T. S. K., and Snyder, S. H.,** Nitric oxide: a physiologic mediator of penile erection, *Science,* 257, 401, 1992.

535. **Finkel, M. S., Oddis, C. V., Jacob, T. D., Watkins, S. C., Hattler, B. G., and Simmons, R. L.,** Negative inotropic effects of cytokines on the heart mediated by nitric oxide, *Science,* 257, 387, 1992.

536. **Kilbourn, R., and Belloni, P.,** Endothelial cells produce nitrogen oxides in response to interferon-gamma, tumour necrosis factor and endotoxin, in *Nitric Oxide From L-Arginine: A Bioregulatory System,* Moncada, S., and Higgs, E. A., Eds., Elsevier, Amsterdam, 1990, chap. 7.

537. **Zhuo, M., Small, S. A., Kandel, E. R., and Hawkins, R. D.,** Nitric oxide and carbon monoxide produce activity-dependent long-term synaptic enhancement in hippocampus, *Science,* 260, 1946, 1993.

538. **Anonymous,** Nitric oxide and glutamate signal transduction, *Neuroscience Facts,* 1, 1, 1990.

539. **Anonymous,** The NO hypothesis, *Neuroscience Facts,* 1, 2, 1990.

540. **Anonymous,** The nitric oxide-calmodulin connection, *Neuroscience Facts,* 1, 2, 1990.

541. **Berridge, M. J.,** Cytosolic Ca^{2+} spiking: mechanisms and functional significance, *Neuroscience Facts,* 3, 73, 1992.

542. **Howard, M., Grimaldi, J. C., Bayan, J. F., Lund, F. E., Santos-Argumedo, L., Parkhouse, R. M. E., Walseth, T. F., and Lee, H. C.,** Formation and hydrolysis of cyclic ADP-ribose catalyzed by lymphocyte antigen CD38, *Science,* 262, 1056, 1993.

543. **Mickelson, J. R., Knudson, C. M., Kennedy, C. F. H., Yang, D.-I., Litterer, L. A., Rempel, W. E., Campbell, K. P., and Louis, C. F.,** Structural and functional correlates of a mutation in the malignant hyperthermia-susceptible pig ryanodine receptor, *FEBS Letters,* 301, 49, 1992.

544. **Mulley, J. C., Kozman, H. M., Phillips, H. A., Gedeon, A. K., McCure, J. A., Iles, D. E., Gregg, R. G., Hogon, K., Couch, F. J., MacLennan, D. H., and Haan, E. A.,** Refined genetic localization for central core disease, *American Journal of Human Genetics,* 52, 398, 1993.

545. **Olken, N. M., Rusche, K. M., Richards, M. K., and Marletta, M. A.,** Inactivation of macrophage nitric oxide synthase activity by N^G-methyl-L-arginine, *Biochemical and Biophysical Research Communications,* 177, 828, 1991.

546. **Feng, S., and Holland, E. C.,** HIV-I tat trans-activation requires the loop sequence within tar, *Nature,* 334, 165, 1988.

547. **Wroblewski, J. T.,** Na^+/Ca^{2+} exchange and glutamate neurotoxicity, *Neuroscience Facts,* 3, 76, 1992.

548. **Verma, A. , Hirsch, D. J., Glatt, C. E., Ronnett, G. V., and Snyder, S. H.,** Carbon monoxide: a putative neural messenger, *Science,* 259, 381, 1993.

549. **Goureau, O., Lepoivre, M., Becquet, F., and Courtois, Y.,** Differential regulation of inducible nitric oxide synthase by fibroblast growth factors and transforming growth factor beta in bovine retinal pigmented epithelial cells: inverse correlation with cellular proliferation, *Proceedings of the National Academy of Sciences of the USA,* 90, 4276, 1993.

550. **Glickman, R. D., and Lam, K.-W.,** Oxidation of ascorbic acid as an indicator of photooxidative stress in the eye, *Photochemistry and Photobiology,* 55, 191, 1992.

551. **Martin, J. B.,** Molecular genetics of neurological diseases, *Science,* 262, 674, 1993.

552. **van del Pol, A. N., Wuarin, J.-P., and Dudek, F. E.,** Glutamate, the dominant excitatory transmitter in neuroendocrine regulation, *Science,* 250, 1276, 1990.

553. **Markert, M., Carnal, B., and Mauel, J.,** Nitric oxide production by activated human neutrophils exposed to sodium azide and hydroxylamine: the role of oxygen radicals, *Biochemical and Biophysical Research Communications,* 199, 1245, 1994.

554. **Louis, J.-C., Magal, E., Takayama, S., and Varon, S.,** CNTF protection of oligodendrocytes against natural and tumor necrosis factor-induced death, *Science,* 259, 689, 1993.

555. **Gross, S. S., Stuehr, D. J., Aisaka, K., Jaffe, E. A., Levi, R., and Griffith, O. W.,** Macrophage and endothelial cell nitric oxide synthesis: cell-type selective inhibition by N^G-aminoarginine, N^G-nitroarginine, and N^G-methylarginine, *Biochemical and Biophysical Research Communications,* 170, 96, 1990.

556. **Hope, B. T., Michael, G. J., Knigge, K. M., and Vincent, S. R.,** Neuronal NADPH diaphorase is a nitric oxide synthase, *Proceedings of the National Academy of Sciences of the USA,* 88, 2811, 1991.

557. **Bredtzen, K., Mandrup-Poulson, T., Nerup, J., Nielsen, J. H., Dinarello, C. A., and Svenson, M.,** Cytotoxicity of human p17 interleukin-1 for pancreatic islets of Langerhans, *Science,* 232, 1545, 1986.

558. **Werner-Felmayer, G., Werner, E. R., Fuchs, D., Hausen, A., Reibnegger, G., and Wachter, H.,** Tetrahydrobiopterin-dependent formation of nitrite and nitrate in murine fibroblasts, *Journal of Experimental Medicine,* 172, 1599, 1990.

559. **Adamson, G. M., and Billings, R. E.,** Cytokine toxicity and induction of NO synthase activity in cultured mouse hepatocytes, *Toxicology and Applied Pharmacology,* 119, 100, 1993.

560. **Meister, A.,** New aspects of glutathione biochemistry and transport: selective alteration of glutathione metabolism, *Federation Proceedings,* 43, 3031, 1984.

561. **Griffith, O. W.,** Glutathione turnover in human erythrocytes. inhibition by buthionine sulfoxime and incorporation of glycine by exchange, *Journal of Biological Chemistry,* 256, 4900, 1981.

562. **Pruett, S. B., Higginbotham, J. N., and Kiel, J. L.,** Quantitative aspects of the feeder cell phenomenon: direct assessment of enhanced cysteine uptake by lymphocytes, *Immunobiology,* 179, 308, 1989.

563. **Herskowitz, K., Bode, B. P., Block, E. R., and Souba, W. W.,** The effects of endotoxin on glutamine transport by pulmonary artery endothelial cells, *Journal of Surgical Research,* 50, 356, 1991.

564. **Boniface, J. J., and Reichert, L. E., Jr.,** Evidence for a novel thioredoxin-like catalytic property of gonadotropic hormones, *Science,* 247, 61, 1990.

565. **Borghetti, A. F., Tramacere, M., Ghiringhelli, P., Severini, A., and Kay, J. E.,** Amino acid transport in pig lymphocytes enhanced activity of transport system ASC following mitogenic stimulation, *Biochimica et Biophysica Acta,* 646, 218, 1981.

566. **Orrenius, S., Ormstad, K., Thor, H., and Jewell, S. A.,** Turnover and functions of glutathione studied with isolated hepatic and renal cells, *Federation Proceedings,* 42, 3177, 1983.

567. **Kavanagh, T. J., Martin, G. M., Livesey, J. C., and Rabinovitch, P. S.,** Direct evidence of intercellular sharing of glutathione via metabolic cooperation, *Journal of Cellular Physiology,* 137, 353, 1988.

568. **Ishii, T., Sugita, Y., and Bannai, S.,** Regulation of glutathione levels in mouse spleen lymphocytes by transport of cysteine, *Journal of Cellular Physiology,* 133, 330, 1987.

569. **Bannai, S., and Ishii, T.,** Formation of sulfhydryl groups in the culture medium by human diploid fibroblasts, *Journal of Cellular Physiology,* 104, 215, 1980.

570. **Bannai, S., and Ishii, T.,** A novel function of glutamine in cell culture: utilization of glutamine for the uptake of cystine in human fibroblasts, *Journal of Cellular Physiology,* 137, 360, 1988.

571. **Meister, A.,** On the enzymology of amino acid transport, *Science,* 180, 33, 1973.

572. **Kouttab, N. M., Mehta, S. R., and Maizel, A. L.,** Human lymphocyte proliferation: requirements for activation and growth, in *Biological Response Modifiers: New Approaches to Disease Intervention,* Torrence, P. F., Ed., Academic Press, Orlando, FL, 1985, chap. 15.

573. **Parker, B. M., McAllister, C. G., and Laux, D. C.,** Lectin-dependent cell-mediated cytotoxicity following *in vitro* culture of normal lymphocytes in medium containing 2-mercaptoethanol, *Immunological Communications,* 11, 387, 1982.

574. **Pruett, S. B., and Kiel, J. L.,** Relationship between oxidative metabolism and thiol production in macrophage-like cell lines, presented at the Federation of American Societies for Experimental Biology 72nd Annual Meeting, Las Vegas, May 1 to 5, 1988, Abstr. 3611.

575. **Henricks, P. A. J., van der Tol, M. E., Thyssen, R. M. W. M., van Asbeck, B. S., and Verhoef, J.,** *Escherichia coli* lipopolysaccharides diminish and enhance cell function of human polymorphonuclear leukocytes, *Infection and Immunity,* 41, 294, 1983.

576. **Raetz, C. R. H.,** Bacterial endotoxins: extraordinary lipids that activate eucaryotic signal transduction, *Journal of Bacteriology,* 175, 5745, 1993.

577. **Rowley, D. A., and Halliwell, B.,** Superoxide-dependent formation of hydroxyl radicals in the presence of thiol compounds, *FEBS Letters,* 138, 33, 1982.

578. **Hampton, R. Y., Golenbock, D. T., and Raetz, C. R. H.,** Lipid A binding sites in membranes of macrophage tumor cells, *Journal of Biological Chemistry,* 263, 14802, 1988.

579. **Cavaillon, J.-M., Fitting, C., Hauttecoeur, B., and Haeffner-Cavaillon, N.,** Inhibition by gangliosides of the specific binding of lipopolysaccharide (LPS) to human monocytes prevents LPS-induced interleukin-1 production, *Cellular Immunology,* 106, 293, 1987.

580. **Wright, S. D., and Jong, M. T. C.,** Adhesion-promoting receptors on human macrophages recognize *Escherichia coli* by binding to lipopolysaccharide, *Journal of Experimental Medicine,* 164, 1876, 1986.

581. **Schwertner, H. A., Robinson, D. K., Kiel, J. L., and Dunham, R. G.,** Effects of microwave fields on amino acid metabolism in mouse macrophage cell cultures, in *Electricity and Magnetism in Biology and Medicine,* Blank, M., Ed., San Francisco Press, San Francisco, 1993, 892.

582. **Pfeifer, R. W., and Irons, R. D.,** Mechanisms of sulfhydryl-dependent immunotoxicity, in *Immunotoxicology and Immunopharmacology,* Dean, J., Luster, M., Munson, A., and Amos, H., Eds., Raven Press, New York, 1985, 255.

583. **Hoffeld, J. T.,** Agents which block membrane lipid peroxidation enhance mouse spleen cell immune activities *in vitro:* relationship to the enhancing activity of 2-mercaptoethanol, *European Journal of Immunology,* 11, 371, 1981.

584. **Arrick, B. A., Nathan, C. F., Griffith, O. W., and Cohn, Z. A.,** Glutathione depletion sensitizes tumor cells to oxidative cytolysis, *Journal of Biological Chemistry,* 257, 1231, 1982.

585. **Baker, M. A., Taylor, Y. C., and Brown, J. M.,** Radiosensitization, thiol oxidation, and inhibition of DNA repair by SR4077, *Radiation Research,* 113, 346, 1988.

586. **Arrick, B. A., Nathan, C. F., and Cohn, Z. A.,** Inhibition of glutathione synthesis augments lysis of murine tumor cells by sulfhydryl-reactive antineoplastics, *Journal of Clinical Investigations,* 71, 258, 1983.

587. **Arrick, B. A., and Nathan, C. F.,** Glutathione metabolism as a determinant of therapeutic efficiency: a review, *Cancer Research,* 44, 4224, 1984.

588. **Mitchell, J. B., Russo, A., Kinsella, T. J., and Glatstein, E.,** Glutathione elevation during thermotolerance induction and thermosensitization by glutathione depletion, *Cancer Research,* 43, 987, 1983.

589. **Deiss, L. P., and Kinchi, A.,** A genetic tool used to identify thioredoxin as a mediator of a growth inhibitory signal, *Science,* 252, 117, 1991.

590. **Harbrecht, B. G., Stadler, J., Billiar, T. R., Ochoa, J. B., Curran, R. D., and Simmons, R. L.,** Inhibiition of glutathione metabolism decreases hepatocyte nitric oxide synthesis, presented at the 75th Annual FASEB Meeting, Atlanta, April 21 to 25, 1991, Abstr. 4364.

591. **Misra, H. P.,** Generation of superoxide free radical during autoxidation of thiols, *Journal of Biological Chemistry,* 249, 2151, 1974.

592. **Abate, C., Patel, L., Rauscher III, F. J., and Curran, T.,** Redox regulation of fos and jun DNA-binding activity in vitro, *Science,* 249, 1157, 1990.

593. **Frankel, A. D., Bredt, D. S., and Pabo, C. O.,** Tat protein from human immunodeficiency virus forms a metal-linked dimer, *Science,* 240, 70, 1988.

594. **Olson, T. S., and Lane, M. D.,** A common mechanism for posttranslational activation of plasma membrane receptors, *FASEB Journal,* 3, 1618, 1989.

595. **Tu, A. T.,** *Venoms: Chemistry and Molecular Biology,* John Wiley & Sons, New York, 1977, chap. 18.

596. **Greenberg, R. N., Dunn, J. A., and Gluerrant, R. L.,** Reduction of the secretory response to *Escherichia coli* heat-stable enterotoxin by thiol and disulfide compounds, *Infection and Immunity,* 41, 174, 1983.

597. **Punjabi, C. J., Laskin, D. L., MacEachen, L., and Laskin, J. D.,** Benzene treatment of mice results in enhanced production of nitric oxide by bone marrow leukocytes, *The Toxicologist,* 13, 183, Abstr. 661, 1993.

598. **Pfeifer, R. W., and Irons, R. D.,** Inhibition of lectin-stimulated lymphocyte agglutination and mitogenesis by hydroquinone: reactivity with intracellular sulfhydryl groups, *Experimental and Molecular Pathology,* 35, 189, 1981.

599. **Fischman, C. M., Udey, M. C., Kurtz, M., and Wedner, H. J.,** Inhibition of lectin-induced lymphocyte activation by 2-cyclohexene-1-one: decreased intracellular glutathione inhibits an early event in the activation sequence, *Journal of Immunology,* 127, 2257, 1981.

600. **Hoffeld, J. T.**, Inhibition of lymphocyte proliferation and antibody production *in vitro* by silica, talc, bentonite or *Corynebacterium parvum*: involvement of peroxidative processes, *European Journal of Immunology*, 13, 364, 1983.

601. **Aune, T. M., and Pierce, C. W.**, Identification and initial characterization of a nonspecific suppressor factor produced by soluble immune response suppressor (SIRS)-treated macrophages, *Journal of Immunology*, 127, 1828, 1981.

602. **Le Boeuf, R. D., Burns, J. N., Bost, K. L., and Blalock, J. E.**, Isolation, purification, and partial characterization of suppressin, a novel inhibitor of cell proliferation, *Journal of Biological Chemistry*, 265, 158, 1990.

603. **Kasukabe, T., Okabe-Kado, J., Honma, Y., and Hozumi, M.**, Purification of a novel growth inhibitory factor for partially differentiated myeloid leukemic cells, *Journal of Biological Chemistry*, 263, 5431, 1988.

604. **Fujiuara, H., and Ellner, J. J.**, Spontaneous production of a suppressor factor by the human macrophage-like cell line U937. I. Suppression of interleukin 1, interleukin 2, and mitogen-induced blastogenesis in mouse thymocytes, *Journal of Immunology*, 136, 181, 1986.

605. **Irons, R. D., Pfeifer, R. W., Aune, T. M., and Pierce, C. W.**, Soluble immune response suppressor (SIRS) inhibits microtubule function in vivo and microtubule assembly *in vitro*, *Journal of Immunology*, 133, 2032, 1984.

606. **Bilzer, M., Krauth-Siegel, R. L., Schirmer, R. H., Akerboom, T. P. M., Sies, H., and Schulz, G. E.**, Interaction of a glutathione S-conjugate with glutathione reductase. Kinetic and X-ray crystallographic studies, *European Journal of Biochemistry*, 138, 373, 1984.

607. **Gelfand, E. W., Cheung, R. K., Mills, G. B., Lee, J. W., Lederman, H. M., Nisbet-Brown, E., and Grinstein, S.**, Analysis of calcium-dependent and calcium-independent signals in triggering human T lymphocytes, in *Primary Immunodeficiency Disease*, Eibl, M. M., and Rosen, F. S., Eds., Elsevier, Amsterdam, 1986, 153.

608. **Nishibe, S., Wahl, M. I., Hernandez-Sotomayor, S. M. T., Tonks, N. K., Rhee, S. G., and Carpenter, G.**, Increase of the catalytic activity of phospholipase C-gamma 1 by tyrosine phosphorylation, *Science*, 250, 1253, 1990.

609. **Scott, W. A., Rouzer, C. A., and Cohn, Z. A.**, Leukotriene C release by macrophages, *Federation Proceedings*, 42, 129, 1983.

510. **Marfat, A., and Corey, E. J.**, Synthesis and structure elucidation of leukotrienes, in *Advances in Prostaglandin, Thromboxane, and Leukotriene Research*, Vol. 14, Pike, J. E., and Morton, D. R., Jr., Eds., Raven Press, New York, 1985, 155.

611. **Monks, T. J., Anders, M. W., Dekant, W., Stevens, J. L., Lau, S. S., and van Bladeren, P. J.**, Glutathione conjugate mediated toxicities, *Toxicology and Applied Pharmacology*, 106, 1, 1990.

612. **Kauffman, F. C.**, Conjugation-deconjugation reactions in drug metabolism and toxicity, *Federation Proceedings*, 46, 2434, 1987.

613. **Anders, M. W.**, Glutathione-dependent bioactivation of xenobiotics, *Lilly Research Laboratories Symposium/Molecular Toxicology*, May, 87, 1990.

614. **Beck, W. T.**, Increase by vinblastine of oxidized glutathione in cultured mammalian cells, *Biochemical Pharmacology*, 29, 2333, 1980.

615. **Calvin, H. I., Medvedovsky, C., and Worgul, B. V.**, Near-total glutathione depletion and age-specific cataracts induced by buthionine sulfoximine in mice, *Science*, 23, 553, 1986.

616. **Gelatt, K. N., Bruss, M., De Costanza, S. M., Noonan, N. E., Das, N. D., and Wolf, E. D.**, Reduced, oxidized, and protein-bound glutathione concentrations in normal and cataractous lenses in the dog, *American Journal of Veterinary Research*, 43, 1215, 1982.

617. **Kelley, S. L., Basu, A., Teicher, B. A., Hacker, M. P., Hamer, D. H., and Lazo, J. S.**, Overexpression of metallothionein confers resistance to anticancer drugs, *Science*, 241, 1813, 1988.

618. **Shalaby, M. R., Espevik, T., Rice, G. C., Ammann, A. J., Figari, I. S., Ranges, G. E., and Palladino, M. A., Jr.,** The involvement of human tumor necrosis factors-alpha and -beta in the mixed lymphocyte reaction, *Journal of Immunology,* 141, 499, 1988.

619. **Raschke, W. C., Baird, S., Ralph, P., and Nakoinz, I.,** Functional macrophage cell lines transformed by Abelson leukemia virus, *Cell,* 15, 261, 1978.

620. **Bernton, E. W., Meltzer, M. S., and Holaday, J. W.,** Suppression of macrophage activation and T-lymphocyte function in hypoprolactinemic mice, *Science,* 239, 401, 1988.

621. **Rimsky, L., Hauber, J., Dukovich, M., Malim, M. H., Langlois, A., Cullen, B. R., and Greene, W. C.,** Functional replacement of the HIV-I rev protein by the HTLV-I rex protein, *Nature,* 335, 738, 1988.

622. **Malim, M. H., Hauber, J., Fenrick, R., and Cullen, B. R.,** Immunodeficiency virus rev trans-activator modulates the expression of the viral regulatory genes, *Nature,* 335, 181, 1988.

623. **Weeks, K. M., Ampe, C., Schultz, S. C., Steitz, T. A., and Crothers, D. M.,** Fragments of the HIV-I TAT protein specifically bind TAR RNA, *Science,* 249, 1281, 1990.

624. **Gatignol, A., Buckler-White, A., Berkhout, B., and Jeang, K.-T.,** Characterization of a human TAR-RNA-binding protein that activates the HIV-I LTR, *Science,* 251, 1597, 1991.

625. **Parker, J. E., Alls, J. L., Holwitt, E., and Kiel, J. L.,** A cell growth factor produced by RAW 264.7 cells after high cell density stress, presented at the annual Meeting of the American Society of Cell Biology, Boston, December 8 to 12, 1991.

626. **Valerie, K., Delers, A., Bruck, C., Thiriart, C., Rosenberg, H., Debouck, C., and Rosenberg, M.,** Activation of human immunodeficiency virus type 1 by DNA damage in human cells, *Nature,* 333, 78, 1988.

627. **Maddox, J.,** Where the AIDS virus hides away, *Nature,* 362, 287, 1993.

628. **Temin, H. M., and Bolognesi, D. P.,** Where has HIV been hiding? *Nature,* 362, 292, 1993.

629. **Embretson, J., Zupancic, M., Ribas, J. L., Burke, A., Racz, P., Tennes-Racz, K., and Haase, A. T.,** Massive covert infection of helper T lymphocytes and macrophages by HIV during the incubation period of AIDS, *Nature,* 362, 359, 1993.

630. **McKeever, P. E., Walsh, G. P., Storrs, E. E., and Balentine, J. D.,** Electron microscopy of peroxidase and acid phosphatase in leprous and uninfected armadillo macrophages: a macrophage subpopulation contains peroxisomes and lacks bacilli, *American Journal of Tropical Medicine and Hygiene,* 27, 1019, 1978.

631. **Nathan, C. F., Kaplan, G., Levis, W. R., Nusrat, A., Witmer, M. D., Sherwin, S. A., Job, C. K., Horowitz, C. R., Steinman, R. M., and Cohn, Z. A.,** Local and systemic effects of intradermal recombinant interferon-gamma in patients with lepromatous leprosy, *The New England Journal of Medicine,* 315, 6, 1986.

632. **Dunn-Coleman, N. S., Smarrelli, J., Jr., and Garrett, R. H.,** Nitrate assimilation in eukaryotic cells, *International Review of Cytology,* 92, 1, 1984.

633. **Miyazaki, J., Juricek, M., Angelis, K., Schnorr, K. M., Kleinhofs, A., and Warner, R. L.,** Characterization and sequence of a novel nitrate reductase from barley, *Molecular General Genetics,* 228, 329, 1991.

634. **Somers, D. A., Kuo, T.-M., Kleinhofs, A., and Warner, R. L.,** Nitrate reductase-deficient mutants in barley. Immunoelectrophoretic characterization, *Plant Physiology,* 71, 145, 1983.

635. **Somers, D. A., Kuo, T.-M., Kleinhofs, A., Warner, R. L., and Oaks, A.,** Synthesis and degradation of barley nitrate reductase, *Plant Physiology,* 72, 949, 1983.

636. **Chaudhry, G. R., and MacGregor, C. H.,** Cytochrome b from *Escherichia coli* nitrate reductase. Its properties and association with the enzyme complex, *Journal of Biological Chemistry,* 258, 5819, 1983.

637. **Dong, X.-R., Li, S. F., and De Moss, J. A.,** Upstream sequence elements required for narL-mediated activation of transcription from the narGHJI promoter of *Escherichia coli,* *Journal of Biological Chemistry,* 267, 14122, 1992.

638. **Stewart, V., Parales, Jr., J., and Merkel, S. M.,** Structure of genes narL and narX of the narL nitrate reductase locus in *Escherichia coli* K-12, *Journal of Bacteriology,* 171, 2229, 1989.

639. **Rabin, R. S., Collins, L. A., and Stewart, V.,** In vivo requirement of integration host factor for nar (nitrate reductase) operon expression in *Escherichia coli* K-12, *Proceedings of the National Academy of Sciences of the USA,* 89, 8701, 1992.

640. **Iobbi, C., Santini, C.-L., Bonnefoy, V., and Giordano, G.,** Biochemical and immunological evidence for a second nitrate reductase in *Escherichia coli* K12, *European Journal of Biochemistry,* 168, 451, 1987.

641. **Siddiqui, R. A., Warnecke-Eberz, U., Hengsberger, A., Schneider, B., Kostka, S., and Friedrich, B.,** Structure and function of a periplasmic nitrate reductase in *Alcaligenes eutrophus* H16, *Journal of Bacteriology,* 175, 5867, 1993.

642. **Lee, H. S., Abdelal, A. H. T., Clark, M. A., and Ingraham, J. L.,** Molecular characterization of nosA, a *Pseudomonas stutzeri* gene encoding an outer membrane protein required to make copper-containing N_2O reductase, *Journal of Bacteriology,* 173, 5406, 1991.

643. **Heiss, B., Frunzke, K., and Zumft, W. G.,** Formation of the N–N bond from nitric oxide by a membrane-bound cytochrome bc complex of nitrate-respiring (denitrifying) *Pseudomonas stutzeri, Journal of Bacteriology,* 171, 3288, 1989.

644. **Jungst, A., Braun, C., and Zumft, W. G.,** Close linkage in *Pseudomonas stutzeri* of the structural genes for respiratory nitrite reductase and nitrous oxide reductase, and other essential genes for denitrification, *Molecular General Genetics,* 225, 241, 1991.

645. **Braun, C., and Zumft, W. G.,** Marker exchange of the structural genes for nitric oxide reductase blocks the denitrification pathway of *Pseudomonas stutzeri* at nitric oxide, *Journal of Biological Chemistry,* 266, 22785, 1991.

646. **Fewson, C. A., and Nicholas, D. J. D.,** Nitrate reductase from *Pseudomonas aeruginosa, Biochimica et Biophysica Acta,* 49, 335, 1961.

647. **Jungst, A., Wakabayashi, S., Matsubara, H., and Zumft, W. G.,** The nirSTBM region coding for cytochrome cd_1-dependent nitrite respiration of *Pseudomonas stutzeri* consists of a cluster of mono-, di-, and tetraheme proteins, *FEBS Letters,* 279, 205, 1991.

648. **Walker, G. C., and Nicholas, D. J. D.,** Nitrite reductase from *Pseudomonas aeruginosa, Biochimica et Biophysica Acta,* 49, 350, 1961.

649. **Braun, C., and Zumft, W. G.,** The structural genes of the nitric oxide reductase complex form *Pseudomonas stutzeri* are part of a 30-kilobase gene cluster for denitrification, *Journal of Bacteriology,* 174, 2394, 1992.

650. **Ye, R. W., Toro-Suarez, I., Tiedje, J. M., and Averill, B. A.,** $H_2{}^{18}O$ isotope exchange studies on the mechanism of reduction of nitric oxide and nitrite to nitrous oxide by denitrifying bacteria. evidence for an electrophilic nitrosyl during reduction of nitric oxide, *Journal of Biological Chemistry,* 266, 12848, 1991.

651. **Kim, C.-H., and Hollocher, T. C.,** Catalysis of nitrosyl transfer reactions by a dissimilatory nitrite reductase (cytochrome c, d_1), *Journal of Biological Chemistry,* 259, 2092, 1984.

652. **Goretski, J., and Hollocher, T. C.,** Catalysis of nitrosyl transfer by denitrifying bacteria is facilitated by nitric oxide, *Biochemical and Biophysical Research Communications,* 175, 901, 1991.

653. **Sumimoto, H., Sakamoto, N., Nozaki, M., Sakaki, Y., Takeshige, K., and Minakami, S.,** Cytochrome b_{558}, a component of the phagocyte NADPH oxidase, is a flavoprotein, *Biochemical Research Communications,* 186, 1368, 1992.

654. **Cannons, A. C., Iida, N., and Solomonson, L. P.,** Expression of a cDNA clone encoding the haem-binding domain of *Chlorella* nitrate reductase, *Biochemistry Journal,* 278, 203, 1991.

655. **Kuo, T. M., Somers, D. A., Kleinhofs, A., and Warner, R. L.,** NADH-nitrate reductase in barley leaves identification and amino acid composition of subunit protein, *Biochimica et Biophysica Acta,* 708, 75, 1982.

656. **Fernandez, E., Schnell, R., Ranum, L. P. W., Hussey, S. C., Silflow, C. D., and Lefebure, P. A.,** Isolation and characterization of the nitrate reductase structural gene of *Chlamydomonas reinhardtii, Proceedings of the National Academy of Sciences of the USA,* 86, 6449, 1989.

657. **Kindle, K. L., Schnell, R. A., Fernandez, E., and Lefebure, P. A.,** Stable nuclear transformation of *Chlamydomonas* using the *Chlamydomonas* gene for nitrate reductase, *Journal of Cell Biology,* 109, 2589, 1989.

658. **Meyer, C., Levin, J. M., Roussel, J.-M., and Rouze, P.,** Mutational and structural analysis of the nitrate reductase heme domain of *Nicotiana plumbaginifolia, Journal of Biological Chemistry,* 266, 20561, 1991.

659. **Hyde, G. E., Crawford, N. M., and Campbell, W. H.,** The sequence of squash NADH:nitrate reductase and its relationship to the sequences of other flavoprotein oxidoreductases. A family of flavoprotein pyridine nucleotide cytochrome reductases, *Journal of Biological Chemistry,* 266, 23542, 1991.

660. **Ben-Shalom, N., Huffaker, R. C., and Rappaport, L.,** Effect of photosynthetic inhibitors and uncouplers of oxidative phosphorylation on nitrate and nitrite reduction in barley leaves, *Plant Physiology,* 71, 63, 1983.

661. **Lillo, C., and Ruoff, P.,** Hysteretic behavior of nitrate reductase. Evidence of an allosteric binding site for reduced pyridine nucleotides, *Journal of Biological Chemistry,* 267, 13456, 1992.

662. **Kiel, J. L., Parker, J. E., Alls, J. L., and Weber, R. A.,** Self-labeling of bacteria with a luminescent polymer, in *Annual International Conference of the IEEE Engineering in Medicine and Biology Society,* Vol. 13, Nagel, J.H., and Smith, W. M., Eds., IEEE, Philadelphia, 1991, 1605.

663. **Kiel, J. L., Alls, J. L., Weber, R. A., and Parker, J. E.,** U.S. Patent 5,156,971, 1992.

664. **Kiel, J. L., Parker, J. E., Holwitt, E. A., and Alls, J. L.,** Aerobic reductive aromatic denitrification, presented at the Symposium on Biodegradation of Nitroaromatic Compounds, Las Vegas, May 22 to 23, 1994, Poster 23.

665. **Mendelson, N. H., and Thwaites, J. J.,** Bacterial macrofibers: multicellular chiral structures, *ASM News,* 59, 25, 1993.

666. **Bruno, J. G., and Kiel, J. L.,** Effect of radio-frequency radiation (RFR) and diazoluminomelanin (DALM) on the growth potential of bacilli, in *Electricity and Magnetism in Biology and Medicine,* Blank, M., Ed., San Francisco Press, San Francisco, 1993, 231.

667. **Collier, S. W., Storm, J. E., and Bronaugh, R. L.,** Reduction of azo dyes during *in vitro* percutaneous absorption, *Toxicology and Applied Pharmacology,* 118, 73, 1993.

668. **Chadwick, R. W., George, S. E., and Clanton, L. D.,** Interaction between Environmental Toxicants, G.I. Microflora, G.I. Enzyme Activity and Chemical Carcinogenesis, RW5793326, Office of Scientific Research, U.S. Air Force, Bolling AFB, Washington, DC, 1991.

669. **Alexander-Caudle, C.,** personal communication.

670. **Metzler, D. E.,** B*iochemistry: The Chemical Reactions of Living Cells.* Academic Press, New York, 1977, chap. 7.

671. **Sabin, A. B.,** HIV vaccination dilemma, *Nature,* 362, 212, 1993.

672. **Weiss, R. A.,** How does HIV cause AIDS? *Science,* 260, 1273, 1993.

673. **Sullenger, B. A., and Cech, T. R.,** Tethering ribozymes to a retroviral packaging signal for destruction of viral RNA, *Science,* 262, 1566, 1993.

INDEX

A

B

Printed and bound by CPI Group (UK) Ltd, Croydon, CR0 4YY

22/10/2024

01777622-0002